**APPLIED MATHEMATICS**

# Applying Geometry to Everyday Life

Erik Richardson

New York

Published in 2017 by Cavendish Square Publishing, LLC
243 5th Avenue, Suite 136, New York, NY 10016

Copyright © 2017 by Cavendish Square Publishing, LLC

First Edition

No part of this publication may be reproduced, stored in a retrieval system, or transmitted in any form or by any means—electronic, mechanical, photocopying, recording, or otherwise—without the prior permission of the copyright owner. Request for permission should be addressed to Permissions, Cavendish Square Publishing, 243 5th Avenue, Suite 136, New York, NY 10016. Tel (877) 980-4450; fax (877) 980-4454.
Website: cavendishsq.com

This publication represents the opinions and views of the author based on his or her personal experience, knowledge, and research. The information in this book serves as a general guide only. The author and publisher have used their best efforts in preparing this book and disclaim liability rising directly or indirectly from the use and application of this book.

CPSIA Compliance Information: Batch #CS16CSQ

All websites were available and accurate when this book was sent to press.

Library of Congress Cataloging-in-Publication Data

Names: Richardson, Erik, author.
Title: Applying geometry to everyday life / Erik Richardson.
Description: New York : Cavendish Square Publishing, [2017] |
Series: Applied mathematics | Includes bibliographical references and index.
Identifiers: LCCN 2016011480 (print) | LCCN 2016017696 (ebook) |
ISBN 9781502619716 (library bound) | ISBN 9781502619723 (ebook)
Subjects: LCSH: Geometry--Juvenile literature.
Classification: LCC QA445.5 .R534 2017 (print) | LCC QA445.5 (ebook) |
DDC 516--dc23
LC record available at https://lccn.loc.gov/2016011480

Editorial Director: David McNamara
Editor: B.J. Best
Copy Editor: Nathan Heidelberger
Art Director: Jeffrey Talbot
Senior Designer: Amy Greenan
Production Assistant: Karol Szymczuk
Photo Researcher: J8 Media

The photographs in this book are used by permission and through the courtesy of: Pigprox/Shutterstock.com, cover; Marco Arment/CC-BY 2.0, 4; Pingebat/Shutterstock.com, 8; Haris Vythoulkas/Shutterstock.com, 9; 10; Euclid/File:P.Oy.I29.jpg/Wikimedia Commons, 10; Filippo Brunelleschi/Perspective drawing for Church of Santo Spirito in Florence/WikiArt, 11; Public Domain/O. Von Corven/Ancientlibraryalex.jpg/Wikimedia Commons, 12; Library of Congress/Arnold Genthe/File:Edna St. Vincent Millay at Mitchell Kennerley's house.tif/Wikimedia Commons, 16; Grant Faint/Photonica/Getty Images, 18; Universal Images Group/Getty Images, 25 (top); Eugene Ivanov/Shutterstock.com, 25 (bottom); Courtesy of Jacqueline de Jonge", 28; Courtesy © George Hart, http://georgehart.com, 35; 3DMAVR/Shutterstock.com, 38; Public Domain/Rishiyur1/Bicycle dimensions.svg/Wikimedia Commons, 41; Chris Cheadle/All Canada Photos/Getty Images, 53; Digital Vision/DigitalVision/Getty Images, 55; Clive Brunskill/Getty Images, 60; Santifc/File:Horizon, Valencia (Spain).JPG/Wikimedia Commons, 62; Kohjiro Kinno/Sports Illustrated/Getty Images, 67; Vlwisconsin (talk)/BackSmithGrind.jpg/Wikimedia Commons, 68; Michale Jenner/ Robert Harding/Getty Images, 70; Paul Prescott/Shutterstock.com, 74; ssitane/E+/Getty Images, 85; James.Pintar/Shutterstock.com, 88; Sherry Yates Young/Shutterstock.com, 97; Microgen/Shutterstock.com, 114; Robert Warren/Photodisc/Getty Images, 117.

Printed in the United States of America

# TABLE OF CONTENTS

| | |
|---|---|
| 5 | Introduction |
| 7 | **ONE The History of Geometry** |
| 19 | **TWO Geometry in Your Everyday Life** |
| 71 | **THREE Geometry in Others' Everyday Lives** |
| 116 | Conclusion |
| 118 | Glossary |
| 121 | Further Information |
| 124 | Bibliography |
| 126 | Index |
| 128 | About the Author |

We can use the same ideas that make us great at playing video games to make us great at other things—like math.

# INTRODUCTION

Research in psychology continues to help us understand that the more we think of something like it's a kind of complex game, the more we're able to keep a healthy perspective about it. That thought process also makes it easier to keep our motivation for improving our performance. Learning about something like math is a great example of that—especially since math itself is one of the features that we add on to things to help make them more and more like a game. Just think about the math that shows up in video games: energy points, gold coins, the amount of the map we've explored, levels, special tools collected, and so on. That goes for role-playing types of games, too, where you are able to build up your character: faster speed, more spells, a stronger energy shield.

Tinkering around with math ideas is kind of like playing a video game in which you are accomplishing different tasks in order to build a video game. Instead of building up your character's speed or getting a better sword, you are building up her ability to accomplish other things in the game. One of those things that helps her do that is thinking about the puzzles and obstacles as part of the game, and the way to do that is to put some math around them. It's OK if it feels like you just wandered into a hedge maze and got lost. Take a couple breaths and retrace your steps.

This process we're engaged in right here is one of the coolest things the human brain is capable of: thinking about thinking. I'm trying to help you see that math is not about one certain kind of puzzle. ("Oh, these are math puzzles, those are language puzzles," and so on.) I'm trying to help you think about the fact that math is like a special set of glasses that your character puts on to add a layer of understanding to *any* of the puzzles.

Go back up and read through that again. It won't take long. Once you're OK with that idea, go on to the next paragraph.

Even within math, geometry is a special set of tools that are quite different from some of the others. Instead of thinking about numbers of things, which is at the heart of other kinds of math, here we are thinking about space and objects within in. When we talk about volume and area and shape and so on, we're talking about not how many there are, but sort of about how much "there-ness" they have.

But hold on, there's one more door to open before you head into the next level of this video game. Sometimes we are actually talking about how much of something *isn't* there. How much emptiness is going on—the carved out part of a swimming pool, the space inside a thing that could be filled up with stuff, or the tension between what's there and what's not there. Whoa!

OK, get your character ready to open the door and start the next level. Be curious. Push the buttons, open the boxes. Pick up the flashing tools on the ground as your character goes by. You don't always know which information or tools will help later in the game.

## ONE

# The History of Geometry

In this first chapter, we will take a rather short walk through some of the key periods of history as geometry moved from culture to culture. We'll highlight a few of the key figures and achievements along the way. This general framework will give us an organizing skeleton onto which we will be able to hook various ideas and examples.

The four major (admittedly simplified) steps will be from the ancient, ancient past in places like Egypt, India, and Babylon, to the ancient (just one ancient) Greeks, to the Muslims, and, finally, to the West from the **Renaissance** and onward.

## Ancient India, Egypt, and Babylon

There is evidence of geometry in parts and pieces stretching back as far as 3000 BCE in the areas of the Indus Valley and ancient Babylon. There were different ideas and principles having to do with angles and lengths as well as some work on calculating areas and volumes. There is little indication of a formal study as would develop under the Greeks. This period of geometry was focused more on practical applications for things like astronomy, surveying, and engineering. These ancient

Here we see the region of the Middle East where Babylonians developed some of the earliest geometry.

collections even contain a version of what would come to be known as the Pythagorean theorem.

Babylonia was the area now largely contained within Iraq. Among the geometric principles they had worked out were ways to calculate the circumference and area of a circle. They also had a dependable formula for the volume of a cylinder, though there were problems when they tried to transform that to cones. There is also evidence that Babylonians were familiar with a version of the Pythagorean theorem (though it wasn't named that yet). There has also been a very recent discovery that the Babylonians also figured out a way to calculate certain things about astronomy.

8   Applying Geometry to Everyday Life

Indian texts from the first millennium BCE are centered on construction of religious altars. As such, there was some interesting work on Pythagorean triples (numbers that can be the integer lengths of sides of a right triangle). Additionally, they seem to have had a way to use Pythagorean triples to find the area of a rectangle for which the hypotenuse is the diagonal.

In the case of Egypt, they had a few methods for finding the approximate area of a circle, including an approximation of π, such as using the fraction 22/7.

## Classical Greece and the Hellenistic period

In the hands of the ancient Greeks, geometry rose up to be treated as the queen of the sciences and began its transformation into a systematic field of mathematics. The Greeks developed the geometry of new forms, including curves, solids, and surfaces (or **planes**). Not only did they expand the content, they also changed the nature of geometry by developing the form that would define the nature of "proof" for a claim for the next two thousand years.

Among the key contributors in this period were Thales of Miletus (635–543 BCE), Pythagoras (582–496 BCE), and Plato (427–347 BCE). Thales and his (likely) student Pythagoras are usually given the credit for the beginnings of **deductive** proof in geometry. It is

Plato was a famous philosopher and mathematician from ancient Greece who thought the whole universe followed patterns of geometry.

The History of Geometry  9

more difficult to say which of their ideas were original and which were borrowed from the Babylonians and the Indians—such as the Pythagorean theorem itself, which I mentioned. Plato took geometry up into philosophy and even built some of his theories about the universe around geometric concepts.

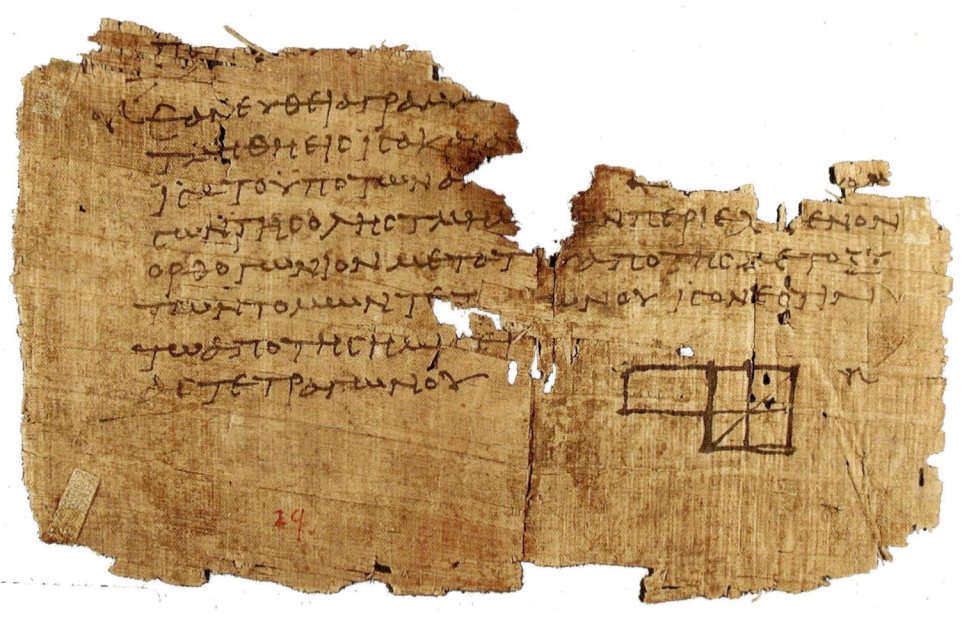

This sample from Euclid's *Elements* allows us to see an early version of one of the most famous math books ever written.

It was with Euclid of Alexandria (325–265 BCE) that things really shifted into a whole different gear, though. He organized a systematic deduction of all the main ideas of geometry. The thirteen-part text he constructed, called *The Elements of Geometry*, was so good that the other texts that had been put together before his fell out of use and out of memory. His *Elements* would be *the* geometry textbook for over a thousand years, and his systematic version still forms the foundation of most beginning high school and college geometry classes.

There was some important work contributed by great thinkers like Archimedes (the Da Vinci of his age, long before there was the actual Da Vinci) and Proclus. But the next great steps forward would really take place in …

## The Islamic period

Some of the work done to move geometry forward during the ninth century CE, which is the next period of real note, involved translating key parts of geometry into numerical operations. This moved beyond the visual **proportional** representations characteristic of the Greek approach. Key players from this period include Al-Mahani, Al-Karaji, and Thabit ibn Qurra. It is the last of those that is of particular interest for his work in fields of analytic geometry, **trigonometry**, and even some ideas of non-Euclidean geometry.

## The Renaissance period

This picture by Brunelleschi was one of the first to use perspective to make drawings and paintings seem more realistic.

Beginning in the tenth century, Europe had taken up and learned from the work that had been done in the Islamic regions. Before that, Europe had struggled through the Dark Ages after the fall of Rome and the Islamic conquest of Alexandria. The primary work done during the subsequent few centuries was centered around the development of perspective as a tool of art for adding realism and perceptual depth to a painting. One mathematician worth noting

The History of Geometry 11

Scholars from different parts of the ancient world would come to study rare texts housed in the library at Alexandria.

12　Applying Geometry to Everyday Life

# A famous library

The Royal Library of Alexandria, Egypt, was created in the third century BCE and stood as probably the greatest center of learning of its time. The library was protected and supported by the rulers of the Ptolemaic dynasty in Egypt. There is some difficulty pinning down all the details because much of the history of the library has been mingled for so long with additional myths. One strong theory is that it was started by one of Aristotle's pupils.

It is hard to say with certainty how many scrolls and codices it held, with some estimating as many four hundred thousand or more. More than just a storehouse working to collect copies of all the world's best writing in every field, the library was a lively place of scholarship and new research—much as one finds at universities now.

It is common enough to hear of "the" burning of the library at Alexandria, but as far as we can tell, there seems to actually have been more than one incident, so dividing up the whole blame is hard to do. Several sources attest to the fact that the last such incident came when the Muslims captured Alexandria in 642 CE and the **caliph** ordered it destroyed. So significant was its loss that it is still referred to as a symbol of lost culture and history.

in this advancement is Brunelleschi, who painted buildings onto mirrors in order to see the way the perspective lines angled. Another is Alberti, who was trained in the science of optics. He wrote a text, *De Pictura*, in about 1435 which laid out the elements of developing perspective in painting.

## The Modern period

As geometry moved into the early seventeenth century, there were two great leaps in mathematical progress. The first was Descartes's creation of analytic geometry, which allowed for algebra to be translated into the coordinate plane and worked with using the visual tools of geometry. The second was the development of calculus by Leibniz and Newton (who both developed it independently of each other). While this is not technically geometry, the implications and applications to geometry helped chart the course of its usefulness and further development since then.

Then, in the eighteenth and nineteenth centuries, several key thinkers developed the growing field that came to be called non-Euclidean geometry. One of the basic rules in Euclid's system had never been grounded in proof the way the others had been, and after several centuries of being unable to do so, a couple of thinkers started thinking outside the box and asked, more or less: "What happens if we kick that **postulate** out and only work with the others?" This rule basically said that area is flat, and two **parallel** lines would never meet. Well, the "what happens" part was that they were able to develop models and theories of geometry that are more useful when dealing with all the parts of the world that do not behave as if they are happening on a flat surface. The two biggest players in this break from tradition were Lobachevsky and Reimann.

While we will not spend time on this area of geometry, I still wanted you to have some familiarity with the main pieces. We can certainly still appreciate the idea and the impact of the sudden historical twist. By way of quick illustration, consider how parallel lines do eventually meet on a curved surface—like the Earth! So different rules and results would follow in terms of understanding how shapes and angles are defined and measured.

## Contemporary period

There are a few other major developments that have taken place within modern geometry, but they do not play the same role of application as Euclidean geometry. They're also very complex, but worth mentioning here. They include topology, field theory, fractals, and abstract algebra. Geometry continues to move on and develop in ways that continue to be more and more fascinating each time. It allows us to use math to get our minds around some new features of the world that we couldn't before.

Please do not misunderstand what this simplified history is showing you. There was still math going on in Greece and India and so on during later centuries. It's not as if the geometry books magically disappeared or something. It just happens that often throughout the history of ideas, one region happens to have the best conditions and best thinkers to make the next big advance before others could.

An overview like this is meant to simplify things enough so that they're easier to hold onto all at once. If you're really interested in a more complex version of how things developed, you should check out more math books or dig around on the Internet to fill in more details. In the meantime, let's jump into an exploration of what all this geometry does when we open the door and let the shapes and formulas and angles scramble out into the world around us.

Edna St. Vincent Millay not only wrote plays and award-winning poetry, but she also wrote one of the first great American operas.

16  Applying Geometry to Everyday Life

# Math in poetry

Sometimes in art we see geometry show up and lend certain impressions and feelings. Other times it is the subject of a work of art. This is true of poetry as well. Sometimes we see math included as one of the writer's tools, but we also occasionally see it as the subject of the poem. A famous example of this is the following poem by Edna St. Vincent Millay. She was a playwright and a Pulitzer Prize–winning poet writing at the early part of the twentieth century. This poem, from 1922, is considered one of her two or three best pieces:

Euclid alone has looked on Beauty bare.
Let all who prate of Beauty hold their peace,
And lay them prone upon the earth and cease
To ponder on themselves, the while they stare
At nothing, intricately drawn nowhere
In shapes of shifting lineage; let geese
Gabble and hiss, but heroes seek release
From dusty bondage into **luminous** air.
O blinding hour, O holy, terrible day,
When first the shaft into his vision shone
Of light anatomized! Euclid alone
Has looked on Beauty bare. Fortunate they
Who, though once only and then but far away,
Have heard her massive sandal set on stone.

A carousel solves a geometric puzzle by folding distances into a smaller space—whether for training knights or amusing people at a theme park.

## TWO

# Geometry in Your Everyday Life

As we try to see those different images—snapshots from history, in a way—let's turn to see some of the images of geometry at work in the present.

Geometry interacts with us in two very different ways. In one way, it is more direct in that geometric shapes and designs clearly stand out for us. This is when geometry is part of the content or even the central subject of a project. In another way, though, it is there as a tool that we use in trying to solve certain puzzles or problems. Geometry allows us to not only be better at our jobs, but also, as you will see, to be better at our various ways of playing and unwinding when we are not working. That includes the ongoing project of coming up with ever-better toys, from amusement park rides to bicycle racing. At the same time, this notion of geometry helping us play includes the ways that an understanding of geometry can help us improve our skill in a sport or pastime.

Here are a variety of different places where circles and angles and areas of planes are coming together like puzzle pieces to help provide the framework of our daily lives and adventures.

# Amusement Park

What better way to start a chapter than a brief imaginary trip to ride the rides (and probably eat too much junk food) at an amusement park? We could spend a lot of time here talking about all the different kinds of math at work, but I'd like to give you a feel for a few of the ways that circles come into play. There are other shapes, of course, but so much of the thrill is from the rides moving in circles. For this part, we can tell ourselves a story about Colleen taking her niece to the park for the afternoon.

## Look at me, I'm a planet

Well, not really, but on a ride like the Scrambler, where each pod of seats spins around at the end of an arm while the whole set of arms is also spinning around, the riders are moving the same way astronomers once thought the planets moved as they orbited around the Earth. My question is: if the pods spin around four times each time the main wheel spins around once, then how far will Colleen and her niece, Bronwyn, travel by the time the ride stops after five times around?

There are two pieces of information we need to know: the radius from the center of the ride out to where the passenger pods are is 30 feet, and the radius of the smaller arm that holds the passenger pod is 2 feet.

So first, let's find the distance they travel when the pod spins around one time:

$$C = 2\pi r_p$$

where:
$C$ = Circumference.
$r_p$ = The radius of the pod.

$$C = 2\pi \times 2$$
$$C = 4\pi$$
$$C = 12.57 \text{ ft}$$

As for the bigger circle:

$$C = 2\pi \times 30$$
$$C = 60\pi$$
$$C = 188.5 \text{ ft}$$

Let's add it all together. They travel four small circles for each big circle, and five times around the big circle. That means:

$$
\begin{array}{rl}
5 \times 188.5 = & 942.5 \text{ ft} \\
20 \times 12.57 = & \underline{251.4 \text{ ft}} \\
& 1{,}193.9 \text{ feet} \approx 400 \text{ yards}
\end{array}
$$

That's going the length of a football field four times! Kind of makes me dizzy just thinking about it.

After taking a break for it to stop feeling like the ground is wobbling, Colleen and Bronwyn hop on the carousel. Bronwyn picks her favorite, a unicorn with a pink harness, which is 8 meters from the center, and Colleen climbs onto a swan, which is closer in at 6 meters from the center. The ride goes on for about 10 minutes, and they go around 30 times. How fast is each one going, and how far do they travel?

Well, we start by figuring out the distance around the two different circles. The formula for the circumference is the same as just above:

$$C = 2\pi r$$

Geometry in Your Everyday Life

So, for Bronwyn:

$$C = 2\pi \times 8$$
$$C = 50.265 \text{ meters}$$

If she goes around 30 times, then that is a total of 1,507.95 meters. And for Colleen:

$$C = 2\pi \times 6$$
$$C = 37.699 \text{ meters}$$

If she goes around 30 times, then that is a total of 1,130.97 meters.

Now we just need to convert those distances to see how fast they are traveling.

$$\frac{3.28084 \text{ ft}}{1 \text{ m}} = \frac{x \text{ ft}}{1,508 \text{ m}}$$

$$x = 4,947.34 \text{ feet in 10 minutes}$$

Let's turn that into an hourly rate via a proportion:

$$4,947.34/10 = x/60$$
$$x = 29,684.04 \text{ feet per hour}$$

There are 5,280 feet in a mile, so Bronwyn is going 29,684/5,280 = 5.62 miles per hour.

Let's see how Colleen is doing by comparison:

$$\frac{3.28084 \text{ ft}}{1 \text{ m}} = \frac{x \text{ ft}}{1,131 \text{ m}}$$

$$x = 3,710.63 \text{ feet in 10 minutes}$$

# The history of carousels

Carousels as we know them actually had their start in **medieval** tournaments. As jousting declined, the knights would instead ride around in circles using their lances to spear small hoops from their holders. This idea had been brought to Europe from the Middle East where the game involved, instead, riders on horseback riding in circles while tossing a ball to each other. This was a game to help train for war.

By the 1700s, these contests had spread throughout France and Italy and were even played by commoners at fairgrounds all over. In Paris, a make-believe carousel was set up for children, complete with wooden horses, and the ride was born. At first, the horses were suspended on chains, and as the central axle turned (by human or by horse), the riders would swing outward slightly—much like a different ride in the modern amusement park.

In the 1800s, the ride transformed into having a floor and horses mounted on poles instead of hanging from above with chains. In 1861, the first steam-powered carousel was introduced by Englishman Thomas Bradshaw at a fair at Aylsham.

The last element of the modern carousel came when Fred Savage, an engineer, added gears and axles to the machinery that would create the up and down motion of the horses. Occasionally you will still find carousels that allow you to try to grab a brass ring as you go around. This is a holdover from the knights trying to get those hoops with their lances.

$$3{,}710.63/10 = x/60$$

$$x = 22{,}263.78 \text{ feet per hour}$$

So Colleen is going 4.22 miles per hour, even though she and Bronwyn are riding the same carousel! **Points** further out on a circle travel faster when that circle is rotated.

## Arts and Crafts

This is an interesting area to consider because it reminds us that geometry, in particular, is reflected in the world in ways that don't always have strictly to do with qualities that can be translated to numbers. Some examples are things like **symmetry**, balance, proportion, and even the impression of texture. That does not even mention the revolution that was created when the use of geometric principles allowed for the discovery of how to create a sense of perspective in art. It is hard for us to imagine now, but at the time it would have been mind blowing how much more realistic art suddenly became. Even more than the first time you saw a movie in 3-D, but a similar kind of experience.

### Famous paintings

Pointillism was a style of art grew that out of the impressionist movement. It emerged in 1886 with Georges Seurat and Paul Signac. Seurat, whose work is displayed on page 25, went on to become one of the most famous painters in this style. The painting is comprised of thousands of individual dots made by a brush! While it might not seem particularly geometric in its composition or use of forms, the remarkable geometric idea brought to life here is the idea that all objects are composed of individual points from the geometer's point of view.

In contrast to the movement of the pointillists breaking everything into tiny geometric points, we see Pablo Picasso and those who came after explore the idea that everything is composed

Seurat's most famous painting, *A Sunday Afternoon on the Isle of La Grande Jatte*, captures the idea of things being made of geometric points.

of geometric planes, another of the foundational building blocks of the geometric worldview. Picasso and the cubists broke the rules of geometric perspective in order to show the full effect of light playing off the different surfaces of a thing. Many of Picasso's paintings show objects from different sides as part of the same image. This movement has continued to influence artists into the present, as shown in this work by contemporary Russian artist Eugene Ivanov.

Geometry in Your Everyday Life   25

## Quilts

Quilting is one of the great traditional crafts that has been passed down generation after generation, and there are still people everywhere who enjoy the combination of artwork, puzzle solving, and useful warmth that results. Here is a sample quilt block pattern to show you how the calculation process starts.

This quilt block is pieced together from sixteen smaller squares. Some of those squares are composed of two equal triangles. If the whole block is 12 inches on each side, determine the amount of blue, red, and yellow fabric needed to make a 48-inch by 48-inch quilt.

Let's start with calculations for one block:

For the blue portion, we see four squares and eight triangles (half squares). We know that each blue square makes up one-fourth of a side of the quilt block, so one-fourth of 12 is 3 inches. Since we're talking about squares, the total area for a blue square is 3 × 3 = 9 square inches. The area of each of the triangles is half that much: (½)$bh$, where $b$ = the base, or 3 inches, and $h$ = the height, or 3 inches. The total area adds up all the blue regions:

$$4 \times 9 \text{ in}^2 = 36 \text{ in}^2$$

26  Applying Geometry to Everyday Life

$$8 \times 4.5 \text{ in}^2 = 36 \text{ in}^2$$

So blue is 72 square inches for each block. We have a 48-inch quilt made of 12-inch blocks. That means there are four blocks across and four blocks down, or sixteen altogether.

$$72 \text{ in}^2 \times 16 = 1{,}152 \text{ in}^2$$

If we translate that to square feet, we get 8 square feet of material needed for the blue parts of the pattern. For red, we see eight triangles, again with $b$ and $h$ = 3 inches.

$$8 \times 4.5 \text{ in}^2 = 36 \text{ in}^2$$
$$36 \text{ in}^2 \times 16 = 576 \text{ in}^2, \text{ and that comes out to 4 square feet}$$

Yellow is four squares, and is pretty simple:

$$4 \times 9 \text{ in}^2 = 36 \text{ in}^2$$
$$36 \text{ in}^2 \times 16 = 576 \text{ in}^2, \text{ so just 4 square feet}$$

Since we have blue fabric for 8 square feet, red for 4 square feet, and yellow for 4 square feet, that ends up with a quilt that is 16 square feet. Remember, our original quilt was 48 inches by 48 inches, which also comes out to 16 square feet. Perfect! This is just one simple example of how geometry plays into a traditional quilt design. Trust me when I say that if you dig around for a while, you will be amazed at some of the amazing artwork at play. Do an Internet search for "award-winning quilts" and dig around for five or ten minutes on the sites that come up. See if you are willing to admit to yourself that it might be a little cooler than you thought.

Let's take a look at a less traditional quilt design. This one was designed by our interview guest, Jacqueline de Jonge. If you look closely, you can see how well she uses things like tessellated triangles to form the circles and pathways. The other thing you should consider is the way the angles and curves seem to balance each other out in the overall effect, as she discusses. In some of her other quilts, the angles have a larger role, and in some, the curves and circles are more prominent, and that difference gives each design its own personality—sort of its own "feel."

## Good taste never goes out of tile

A second area of arts and crafts that takes up geometry in some interesting ways is the use of tessellation. This shows up in a wide range of arts from quilting to woodworking to the clever work of artists like M. C. Escher, and in the intricate ceramic tile work used in architectural features throughout the Middle East.

Tessellation is taking numerous shapes and fitting them together to completely cover a surface. The most basic example to illustrate would be a checkerboard. There we have used the repeated pattern of squares, which fit together with no gaps and no overlaps. They cover the board completely.

There are several interesting facts that we are able to generalize about geometric shapes and tessellation. For instance, **congruent** copies of any given quadrilateral will always tessellate. **Trapezoids**, parallelograms, kites, you name it. It might puzzle you to know that this is why any triangle will tessellate with copies of itself as well. The same is true for any hexagons. Why do you think that is? Put your finger down here to hold your place and think about it for a minute before reading on. Are you back? Well, two copies of any triangle can always be placed edge to edge to form a quadrilateral, and all quadrilaterals tessellate. You should test that out. They might fit together to form a square or a rectangle, or they might fit together to form a parallelogram. That is what equilateral triangles do, for instance. That doesn't answer the hexagon case though, because not all hexagons can be formed from six congruent triangles. I mention this because that is a very common first guess. Rather than just give away the answer this time, I will give you a clue: it is the opposite of how we solved the triangles question.

Of course, as seen in the example on the left, some tessellations are created with a combination of shapes, not merely congruent copies of just one. You could go through a few rounds of trial and error to test out which kinds do and do not work if you're feeling adventurous. As it turns out, we know that in order for shapes to tessellate on a plane, the sum of the angles meeting at a **vertex** (think of it as the intersection) must add up to 360 degrees.

Geometry in Your Everyday Life 29

# Meet Jacqueline de Jonge

Please meet Jacqueline de Jonge, an award-winning quilter from the Netherlands. She has won the Color Trophy for the Open European Quilt Championship three times, and won the Hoffman Challenge for best paper piecing project with her quilt *Celtic Fantasy* in 2013.

*Thank you for joining us Jacqueline. So, tell us a little about your quilting history.*

I started in 1978. I saw a quilt in a magazine, and I really wanted to give it a try. Since then ... well, I guess I'm sort of addicted to quilting.

*What inspired you to use such strong, clean geometric shapes in your designs?*

I always like to challenge myself and I love to work with symmetrical shapes, but there has to be a balance in shapes, curves, colors, and movement in each quilt. I always was good with math and the challenge to find a way to do something difficult is really fun.

*Is there a kind of math problem or application that you like using more than others when you are planning out the dimensions and composition?*

No. I always try to draw a new design onto a piece of wallpaper. Just to see what it looks like and to fill places and corners. After that we use AUTOCAD to turn the drawing into a workable pattern. From my perspective, quilting and designing are more a matter of intuition than planning. I love to design and to bring the circles on my paper to life. If I'm too focused on a design, it won't work for me.

*What do you think helped you engage well with math in school, or what would have helped more?*

I think I had a wonderful teacher. He explained everything in such a logical and easy, almost simple way that I love math.

*Are there other key ways that math figures into being an award-winning quilter than just in the design process?*

I think to be in balance with yourself and always to try to get the best out of yourself. And always say to yourself, when there is something difficult to face, "Think positive: this is a challenge, but I can do this." If you say "This is too difficult, I can't do this," you won't succeed.

*Thanks for sharing your perspective with us, and thanks for all those beautiful quilts you posted on your website at www.becolourful.com.*

Let's test that on a couple cases. The pattern of tiles shown in the image on page 29 has octagons and squares coming together to form vertices. The formula to find the measure of the interior angles of a regular polygon is:

$$A_i = \frac{180(n-2)}{n}$$

where:

$A_i$ = Interior angle.
$n$ = The number of sides of the polygon.

So, for the octagon, we find:

$$A_i = \frac{180 \times (8-2)}{8}$$

$$A_i = \frac{1{,}080}{8}$$

$$A_i = 135 \text{ degrees}$$

And we know the angle of a square is 90 degrees. That means at each of those vertices we have: (2 × 135) + 90 = 360 degrees.

What about the traditional pattern from a soccer ball with a pentagon surrounded by hexagons? Would that tessellate on a plane just as it does on a sphere? Well, each vertex would have a corner of a pentagon and two corners from hexagons, so let's see.

# My favorite M.C.

Few people have played as much or been so clever with tessellation and perspective as M. C. Escher, a famous artist who lived from 1898 to 1972. Escher is known all over the world for his inventive drawings. There's a drawing of a waterfall which seems to run back uphill and start over again (*Waterfall*), a picture of the two hands drawing each other (*Drawing Hands*), and even patterned tessellations using some uncommon shapes to replicate—lizards, horses, and even seahorses.

In addition to his famous drawings, he also illustrated books and designed some postage stamps and tapestries. His particular fascination with tessellation arose from his visit to the Alhambra, the fourteenth-century **Moorish** castle in Spain. Many of the walls there are covered with sculpted tile tessellations or with overlain material to create the same visual effect.

His work, bending and distorting the way viewers expect planes to behave, created interesting opportunities for optical illusions. He played around with ways to represent infinity and, at the same time, did some math research on his own. That interest in mathematics, even though he had no formal training in the field, helps to explain why so much of his work was built with geometric shapes and properties. It also helps to explain his tinkering with the limits of trying to represent 3-D figures on a 2-D piece of paper.

For the pentagon, we get:

$$A_i = \frac{180 \times (5 - 2)}{5}$$

$$A_i = \frac{540}{8}$$

$$A_i = 108 \text{ degrees}$$

For the hexagons, the angles are:

$$A_i = \frac{180 \times (6 - 2)}{6}$$

$$A_i = \frac{720}{8}$$

$$A_i = 120 \text{ degrees}$$

That means we would have (2 × 120) + 108 = 348 degrees. So no, they would not tessellate.

## Sculpture

While quilting and tiling technically produce 3-D objects, the artwork itself is still, essentially, a 2-D format. This seems a very limited sample given the fact that geometry has so much to do with fully embodied three-dimensional shapes out in the real world. To help explore this further facet of geometry in art, then, it seems fitting to spend a few minutes talking about sculpture.

One does not have to search very long on the Internet to find any number of sculptures in varying degrees of geometric clarity.

These sculptures range from collections of simple forms—cubes, spheres, etc.—to subtler, more complex pieces. They might be more like a work of abstract art, but still clearly constructed in the mind by means of planes, angles, and so on.

Below is an interesting piece of contemporary sculpture by the artist and engineer George W. Hart. There are so many things to appreciate here—the combination of spheres with the linear, angular **polyhedral** structure, the balance between organic and inorganic, even the tension between art and useful everyday objects like forks.

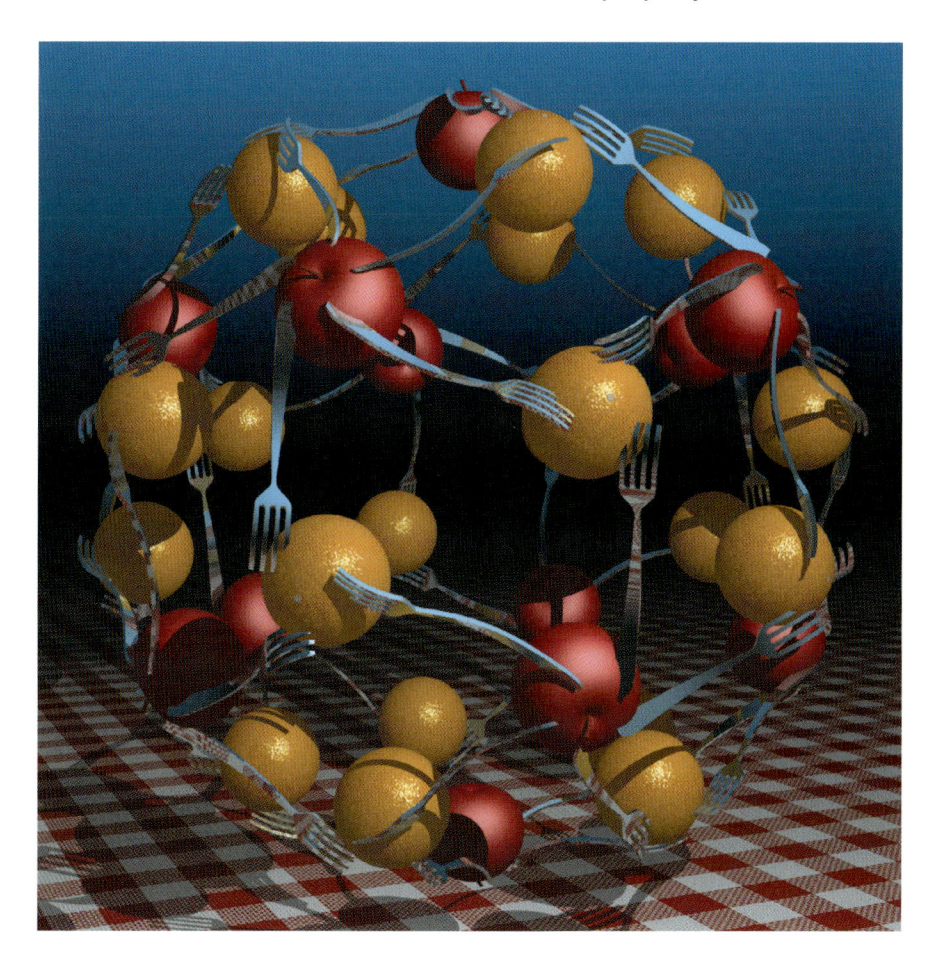

# Meet George W. Hart

Please meet George W. Hart, a professor of engineering at Stony Brook University, a sculptor with works on exhibit around the world, and one of the cofounders of the Museum of Mathematics in New York City.

*So tell us, George, how long have you been sculpting?*

I've been making geometric things since I was a kid. They've just gotten fancier over time.

*What are some different ways you tinker around with math and shapes when you are coming up with the ideas for your pieces?*

I'm always playing with ideas, sometimes with paper or physical models, sometimes by programming to generate computer renderings. When I find something I love, I think of ways to make it in a permanent material.

*What inspired you to use such strong, clean geometric shapes in your designs—was it inspiration from someone? Something about how they create movement?*

I follow my geometric aesthetic. I don't think I can explain it in words. I can just show my sculpture.

*Is there a kind of math problem or application that you like using more than others when you are planning out the dimensions and composition?*

36   Applying Geometry to Everyday Life

Many branches of mathematics inspire me. I often use a lot of trigonometry and linear algebra when working out the details of how to construct a design.

*What do you think helped you engage well with math in school?*

I enjoyed reading a lot of math books. Martin Gardner was a great early inspiration. Nowadays there are videos online one can start with, but there is no substitute for in-depth reading and thinking.

*Are there other key ways that math figures into being a sculptor than just in the design process?*

Doing math teaches people how to patiently explore ideas and persevere to solve challenging problems. Those skills are useful in any field. In sculpture, there are many difficulties of design, fabrication, assembly, etc., which benefit from general problem-solving skills.

*Any interesting or colorful question I didn't think to ask but should have?*

What can we do to get the next generation of people involved in mathematical art?

*Thanks for your perspective, George. I can definitely see why you've been invited to put on workshops at places like Princeton University, TEDx at Wellesley, and in Kyoto and Venice. To see more of his amazing work, go to www.georgehart.com.*

# Machines

## The more things chain, the more they stay the same

When it comes to complex machines, bikes have to be high on the list of all-time favorites. The sense of freedom, the opportunities for reckless danger—bikes expand the mind more in one single dose than almost any other machine I can think of. Isn't it funny that of all the things we have thought about our bikes and done with our bikes, few of us ever stopped to think about all the math going on? I thought it would be fun to just look at two closely related pieces of that to show how math is at play just under the surface.

The first example to point to is the chain. That's the heart of a bike. That's what turns our churning, burning little legs into speed and motion much faster than normal running.

Let's look at how the length of the chain is determined by the size and relationship of the **sprockets**. Sprockets are the toothed metal wheels—gears—that the chain wraps around at both ends. The length between sprockets is relatively easy; it is the length that goes around the curves that forces us to grab for the geometry we have stored away in our heads somewhere.

Take this diagram, for instance:

If two sprockets were the same size, then the chain would start and end its curvature at the vertical axis and the span around the back would equal 180 degrees. However, most of the time, these two sprockets are of differing sizes to allow us an increase in speed, which we'll talk about in a few minutes. But here, let the front sprocket have circumference (*C*) of 22 inches and the back sprocket have *C* of 8 inches.

In this case, the arc of the circle the chain wraps around the small sprocket is 164 degrees. To find how much of the total 8-inch circumference that represents, it's proportion time:

$$\frac{164}{360} = \frac{x}{8}$$

We cross-multiply to get:

$$164 \times 8 = 360x$$

Divide both sides by 360:

$$3.64 \text{ inches} = x$$

In the case of the big sprocket, the portion of the circumference wrapped by the chain is all of it *except* for 164 degrees, so 360 – 164 = 196 degrees.

$$\frac{196}{360} = \frac{x}{22}$$

Again, we go crazy with some cross-multiplying to get:

$$196 \times 22 = 360x$$

Divide both sides by 360:

$$11.978 \text{ inches} = x$$

If the top and bottom chain lengths between the sprockets are both 18 inches, we can add up the sections to get a total length of 18 + 18 + 3.64 + 11.978 = 51.62 inches.

The setup is a little more complicated on a bike with multiple gears because the chain has to go around the extra machinery that moves the chain gradually from one sprocket wheel to the next. The relationship between the circumferences of the two different gears creates a change in whether you are making the back wheel go around faster or slower relative to your pedaling rate.

Think about our example above. Each time you turn the 22-inch sprocket wheel, just think about how many times the 8-inch sprocket wheel in the back would turn. If the sprocket wheels were the same size, what do you think would happen? Right, they'd be the same. So you can see that changing the relative circumferences of the two gears would affect the relationship between how much distance is traveled (how many more times the back wheel turns) with each pedal.

There are a number of other ways that geometry comes into play with respect to good cycling.

In the diagram on the opposite page, we see a number of the key angles that play a role in the balance between maneuverability (how easy the bike is to steer) and stability. One of those is the rake, which is a measure of how far the center of the wheel is ahead of the line in the diagram that shows the steering axis (up through the neck where the handlebars connect to the frame). Another one is called the trail, which measures how much of the section of tire in contact with the road is behind the steering axis. These are helpful, for instance, in understanding that as the trail goes up—which would mean the rake is greater—stability would go up because more of the tire is behind the steering axis. However, this is a trade-off, because that also means the bike might be a little more sluggish for maneuvering. Mountain bikes, for example, would have a different trail than racing bikes, and both would be different than a touring bike (which has the shortest trail, because maneuverability is more important than high-speed stability).

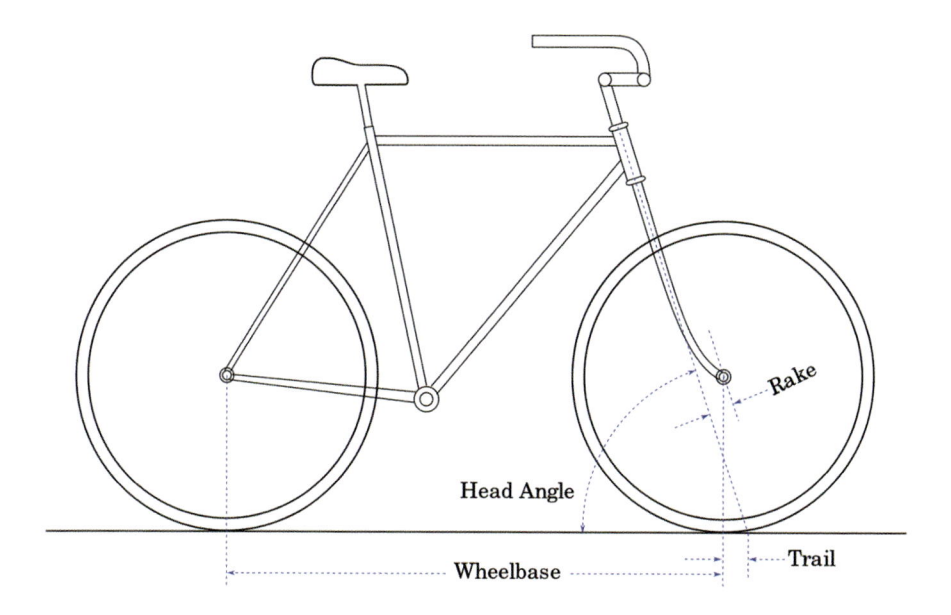

On the other hand, wheelbase, which is the distance between the centers of the front and back wheels, has an impact on a given rider's weight distribution. A general target guideline would be to have 45 percent of weight distributed onto the front wheel and 55 percent on the back wheel. If there is too much weight toward the back, it can be hard to climb because the front wheel wants to pop up.

## Hot air rises, so why call something a heat sink?

Inside electronics, especially computers, there is a lot of heat generated by the electricity passing through the circuits. This is the same principle as why a lightbulb heats up, but to a smaller degree for each wire inside your computer. In order to keep that heat from melting or burning fragile parts inside the circus of circuits, it has to be sent somewhere besides the circuit wiring. The computer has two solutions for that: one is to use a fan to push the hot air out (which draws cooler air in). The other is to absorb the heat into something safe and then let it slowly transfer away from there into the air, instead of being absorbed by the tiny, fragile wiring.

To answer the question: yes, hot air does rise. But these heat absorbers are called heat sinks because they let heat sink into them. That's only part of the coolness, though. (Ha! Heat, coolness ...) What's really interesting is how geometry helps us understand why heat sinks look so funky.

The general principle you have to know is that the larger the surface area, the faster something can cool off when exposed to a cooler thing (like air). The more surface area touching, the faster the energy is transferred between them. Let's see how that is used to improve the design of heat sinks in computers.

Let's start with a nice piece of metal (which is really good at taking in and giving off heat).

If it is 2 inches thick with dimensions of 3 inches by 3 inches for length and width, then we can find its surface area:

$$SA = lw$$

where:

$SA$ = Surface area.
$l$ = Length of a side.
$w$ = Width of a side.

So, the surface area for the top is:

$$SA_t = 3 \times 3$$
$$SA_t = 9 \text{ in}^2$$

The surface area for the bottom is the same, as it has the same dimensions. For the sides:

$$SA_s = 3 \times 2$$
$$SA_s = 6 \text{ in}^2$$

And for each of the two ends:

$$SA_e = 3 \times 2$$
$$SA_e = 6 \text{ in}^2$$

Then we can add them all together:

$$\begin{array}{r} 18 \; (9 \times 2) \text{ top/bottom} \\ 12 \; (6 \times 2) \text{ sides} \\ +12 \; (6 \times 2) \text{ ends} \\ \hline 42 \text{ in}^2 \end{array}$$

Now let's compare that with the measurement after we change it just a little bit. Instead of a block with a flat surface, we'll start simple with just a few grooves added in, so that it looks like this, instead:

Finding the surface area now is going to be a bit trickier. We can treat the bottom like a separate piece, with a height of 1 inch, and we'll leave its top off, because that will be a separate calculation. The length and width are still both 3 inches. That gives us a starting surface area of

$$\underbrace{(3 \times 3)}_{\text{bottom}} + \underbrace{2 \times (3 \times 1)}_{\text{sides}} + \underbrace{2 \times (3 \times 1)}_{\text{ends}} = 21 \text{ in}^2$$

The surface area for the top will include the sides of the triangles where they meet the edge of the block, right? That's going to be a base of 1 and a height of 1, so:

$$A = \tfrac{1}{2}bh$$
$$A = \tfrac{1}{2} \times 1 \times 1$$
$$A = 0.5 \text{ in}^2$$

There will be 3 on this side and 3 on the other side, so:

$$A = 6 \times 0.5 = 3 \text{ in}^2$$

Next, we have the area of the sloped face, which makes a rectangle. We know that it is 3 inches across, but we need Pythagoras's help to figure out how long it is.

$$a^2 + b^2 = c^2$$
$$1^2 + 1^2 = c^2$$
$$2 = c^2$$
$$\sqrt{2} = c$$
$$1.414 = c$$

So the area for each of those sloped rectangles is going to be:

$$A = 3 \times 1.414$$
$$A = 4.242 \text{ in}^2$$

There will be 3 of these, so:

$$A = 3 \times 4.242 = 12.73 \text{ in}^2$$

That just leaves us to find the surface area of the straight backs of the ridges. Each of those is 1 inch high and 3 inches wide, which is simple enough:

$$A = 3 \times 1$$
$$A = 3 \text{ in}^2$$

There will also be 3 of these, so:

$$A = 3 \times 3 = 9 \text{ in}^2$$

When we add these all together, we see that the new surface area created is:

$$21 + 3 + 12.73 + 9 = 45.73 \text{ in}^2$$

From 42 up to 45.73 means we added on an additional 8.9 percent of surface area. In fact, if you look at most heat sinks nowadays, you will see that they have done much more than this. They have turned into a collection of tall, skinny pyramids or cones. To appreciate that, imagine that we now went back and instead of two cuts going across, we also had two cuts going the other way, like a tic-tac-toe board. And then what if we did a bunch of those cuts?

You can appreciate how this strange little feature of geometry and hippopotenuses (it's really hypotenuse, but that's just funnier to say) gives us a little bit of superpowers so we can build better computers. Otherwise, computers would have to run much slower so they wouldn't overheat and melt their circuitry.

## Sports and Recreation

Very often we don't notice how much of our lives and our thinking is about things that aren't even there—the future, things we didn't do, places from books that aren't even real places, and so on. You might even say an idea about how much of life revolves around things that aren't there isn't there.

### Air ball

We see basketballs and footballs and soccer balls and such flying through the air all the time. Have you ever stopped to think about the fact that it's actually a blob of air, wrapped in a piece of material, being thrown through the air? It sounds a little weird when you think about it that way. It does allow for some interesting questions, though, like, "How big is the blob of air we're throwing around and shooting through the hoop?"

Well, a regulation NBA basketball has a circumference of 29.5 inches (while the NCAA, the collegiate sports governing association, sets theirs at 30 inches). In order to calculate the volume, though, we need to work backward to find the radius ($r$). We know that the formula for circumference ($C$) is:

$$C = 2\pi r$$

We can plug in the information given and solve for $r$.

$$29.5 = 2\pi r$$

Divide both sides by 2:

$$14.75 = \pi r$$

Divide both sides by $\pi$:

$$4.695 \text{ in} = r$$

OK, now we can plug that $r$ into the formula for the volume of a sphere:

$$V = (4/3)\pi r^3$$
$$V = (4/3)\pi \times 4.695^3$$
$$V = (4/3)\pi \times 103.492$$
$$V = 137.989\pi$$
$$V = 433.506 \text{ in}^3$$

By NBA regulation, the air inside a basketball needs to be between 7.5 and 8.5 pounds per square inch of pressure. That translates in an interesting way because it would be like taking extra air from the room around us and squishing it down into the basketball, so the air inside the basketball is pushed in tighter than the air around it. That's why a basketball has to be strong enough not to pop or bubble out of shape from the air inside pushing harder than the air outside.

## How good of a pack could a backpacker pack if a backpacker could pack packs?

Here is an interesting puzzle that requires us not only to think through geometry to come up with a good answer, but requires us to compare two different geometry solutions. Let's imagine that our friend Brendan loves to go camping on those rare days he can break away from his bakery for a couple days at a time. This year, he has decided to retire the beat-up, hand-me-down backpack he's been dragging along and invest in a new lightweight pack. He does his comparison shopping and has come up with the following chart of what he feels are the two most important **variables** on a ranking scale of five:

| Pack Brand and Model | Durability | Comfort |
|---|---|---|
| Super Trek Yukon | 5 | 1 |
| Avatar OffTrail | 3 | 4 |
| Camptastic Ultra III | 4 | 2 |
| Hikecules Bearpaw | 3 | 3 |

He now has to decide how best to analyze this comparison chart, and he realizes that the one that graphs farthest from 0 is probably the best. So he summons up the ghost of Pythagoras and runs his calculation with durability as the x-axis variable and comfort as the y-axis variable. Here's the calculation for the SuperTrek Yukon:

$$a^2 + b^2 = c^2$$
$$5^2 + 1^2 = c^2$$
$$25 + 1 = c^2$$
$$26 = c^2$$
$$5.1 = c$$

48  Applying Geometry to Everyday Life

So, the Yukon has a hypotenuse of 5.1, which means that is how far it is from (0,0). If we do the same for the others, we find the following results:

| Super Trek Yukon | 5.1 |
| --- | --- |
| Avatar OffTrail | 5.0 |
| Camptastic Ultra III | 4.5 |
| Hikecules Bearpaw | 4.24 |

According to this method, it looks like the SuperTrek Yukon comes out on top. Let's think about this for a minute, though. Does poor Brendan want to end up with a pack that has the lowest rating for comfort? Two days of hiking is a lot of discomfort. What is more, it is the pack that will last the longest, so he'll be uncomfortable for years!

Let's try an alternative approach. Maybe instead of distance from (0,0) as he first thought, let's look at total area of the graph represented by each point.

| | $A = lw$ |
| --- | --- |
| Super Trek Yukon | $A = 5 \times 1 = 5$ units$^2$ |
| Avatar OffTrail | $A = 3 \times 4 = 12$ units$^2$ |
| Camptastic Ultra III | $A = 4 \times 2 = 8$ units$^2$ |
| Hikecules Bearpaw | $A = 3 \times 3 = 9$ units$^2$ |

Geometry in Your Everyday Life

Under this solution strategy, then, it is the Avatar OffTrail that comes in with the highest score. Brendan is probably going to be a lot better off with that choice, given his two main concerns, because he has not traded pretty much all of one factor just to get a higher score overall.

## Bad puns just keel me

Even when talking about something like recreational boating, geometry comes into play. Many times, certain kinds of boats have a maximum **capacity** label attached, but in the event that a boat doesn't, the limitation is determined by:

$$M = \frac{lw}{15}$$

where:

$M$ = Maximum capacity (the largest number of people the boat can safely hold).
$l$ = Length of the boat in feet.
$w$ = Width of the boat in feet.

Thus, a boat with a length of 25 feet and a width of 8.5 feet would have a capacity of:

$$M = \frac{25 \times 8.5}{15}$$

$$M = \frac{212.5}{15} = 14.16$$

So the safety limit for this boat would be 14 people (and maybe a small dog for the 0.16).

A second place where geometry comes into play has to do with measuring the keel of a sailboat. The keel is a large, heavy fin on the bottom of the boat that helps keep the boat upright. There are several important dimensions that factor into decisions about the keel—especially for boats that engage in any level of racing.

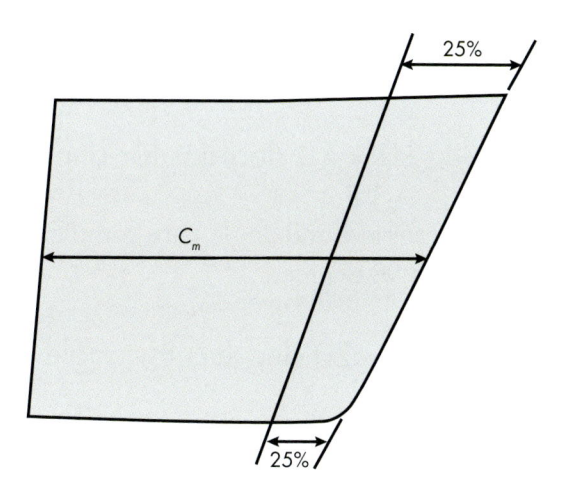

One of these dimensions is something called the mid-chord. This is the measurement from front to back of the keel at a point that is midway between the two bases of the trapezoid.

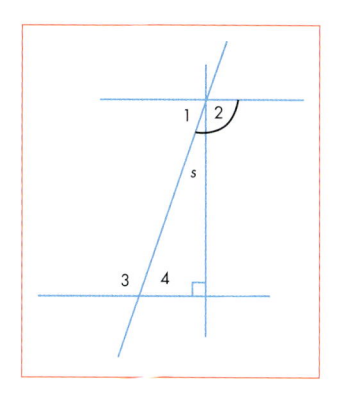

A second dimension would be something called the sweepback angle. This is the measurement for how much the leading edge of the keel is tilted back away from vertical. Because of the curvature and molding at the point where the keel attaches to the hull, we can use a little geometry to figure out what the sweepback angle is based on a few measurements at the mid-chord. As you can see, the

Geometry in Your Everyday Life  51

line for calculating is drawn from a point 25 percent from the edge of the top base (called the root cord) to a point 25 percent from the edge of the bottom base (called the tip chord).

So, if we know that the chords are parallel to each other, and we also know the measure of angle 3 is 110 degrees, how can we use that to find the sweepback angle ($s$)?

Well, we know from postulates of geometry that angle 3 and angle 1 are supplementary, which means they add up to 180 degrees.

$$180 - 110 = 70 \text{ degrees for angle 1}$$

We can also see angle 1 and angle $s$ are complementary, which means they add up to 90 degrees.

$$90 - 70 = 20 \text{ degrees for angle } s$$

Alternatively, we could also have taken the path that angle 3 and angle 2 are congruent (which we know from geometry postulates: given two parallel lines, alternate exterior angles are congruent). That would allow us to use:

$$110 - 90 = 20 \text{ degrees for } s$$

Knowing this sweepback angle, then, allows for improved calculations as we manage variables in the course of a race.

## Is this the wheel life?

Sailors and cyclists aren't the only ones who love the thrill of going fast. Geometry helps us figure out how to arrange a racetrack so that we can run a fair race. If you think about it for a minute, it makes sense that if one of us is running around a big circle and one of us running around a smaller circle, then the small-circle racer has a shorter circumference to travel. Hardly fair. However, putting

These wheelchair racers remind us that geometry is sometimes a tool to help challenge ourselves and see how we measure against the competition.

big curves at the end of racetracks was discovered in ancient days to allow the athletes to keep going full speed but still be close enough for the crowd to watch the whole race.

If we look at the layout of an Olympic-style track, we see two straightaways and two big curved ends. We can use geometry to figure out just how much longer each lane gets. The standards for an Olympic-size track are for the inside line of lane 1 (the innermost lane) to be 400 meters long. This consists of two straightaways of 84.39 meters each and two semicircles of 115.61 meters each. If we know that the radius for the inside line on the lane 1 curves is 36.825 meters, and that the lanes are 1.2 meters wide, then what

Geometry in Your Everyday Life    53

would be the length of the inside line for lane 2? (It's the same line as the outside of lane 1.) Well, let's set up our circumference equation and make the adjustments.

$$C = 2\pi r$$

where:

$C$ = Circumference.
$r$ = Radius.

$$C = 2\pi \times (36.825 + 1.2)$$
$$C = 2\pi \times 38.025$$
$$C = \pi \times 76.05$$
$$C = 238.918 \text{ meters}$$

Wow, that seems *way* bigger than the radius of the semicircle for lane 1. It was only 115.61 meters. That difference is a signal for us to step back for a second and think about what we're doing. Are we sure our numbers are comparing apples to apples? Our number calculates the distance around a circle. Was the 115 number the distance around a circle? No. It was just a semicircle. Now the track has two of these, right? Ours already includes two. So we just need to double the one we started with:

$$2 \times 115.61 = 231.22 \text{ meters}$$

So, in a race that goes all the way around the track one time, how much farther would the person in lane 2 have to travel? About 7.7 meters. If you have ever noticed, that's why they have runners start the long races from different spots—some ahead of others.

## Having a fit in the middle of the store

We talked a bit about the mechanics of bicycles, but there's also a fascinating bundle of geometry that comes up when we talk about the cyclists themselves and the process of analyzing their riding to help improve training and race performance.

In talking with Brent Emery, a competitive cyclist who runs one of the best bike fit programs in the country, he helped to explain the very different ways that a person's physical geometry can influence good decisions about the proper bike size and set up, the best positioning, and so on.

A good fitting starts from a basic template based on the style of riding the person is training for—recreational riding, mountain biking, triathlons, or road racing (like the Tour de France). However, each person is slightly different on any number of variables: length of upper leg, length of arm relative to length of torso, stiffness

For cyclists, even slight changes in angle or arc of different body parts will shift their efficiency and change their speed.

or flexibility level in the lower back, and about a hundred other possible measurements. On top of that, he explained that a given rider's body will change from year to year (or every few years).

During a careful bike fitting and riding analysis, you bring your bike in and place it on a specialized training stand. Then the readouts from the stand (pressure, symmetry of pedaling motion, weight distribution, etc.) are combined with measurements from a laser readout of your pedaling mechanics to help look for places where the angles are too **acute** or **obtuse**, where circular **momentum** is lost, and so on.

In addition to the complex in-store fit analysis, there are some even higher-tech elements that can come into play for the serious cyclist. Even for a former Olympic competitor like Emery—who grew up working in his parents' bike shop, which he now runs—going down to the wind tunnel at Indianapolis still helped him make some slight adjustments to his racing posture and shave some time off his races.

## I'm getting the hang of this

If we switch over from zipping along over the water and the ground to zipping along through the air on a hang glider, we find that triangles are our friends there as well. Consider that in the design of one type of hang glider, the sail (the big, fabric part) is formed by joining two **isosceles** triangles side to side.

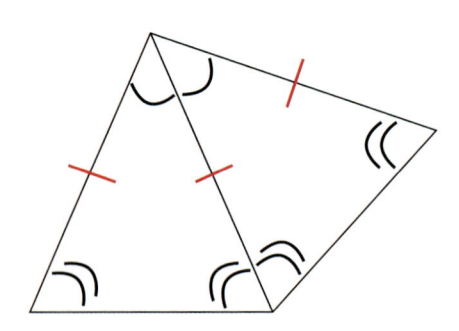

The dimensions have to be such that the nose makes a 90 degree angle. In order to do that, what should the other angles be on the two pieces of fabric before they are joined?

Well, we start with the fact that the sum of the interior angles of a triangle equals 180 degrees. Since the two at the tip come together to make 90 degrees, each must be 45 degrees. That means the other two angles on each triangle add up to:

180 – 45 = 135 degrees

Dividing by two gives us:

135/2 = 67.5 degrees for the other angles

Geometry comes in handy not only for designing the sail, but for looking at the placement of the triangle control bar on a hang glider as well. For this one, it's going to be simpler to use a slightly different hang glider shape which forms one triangle.

In a standard hang glider, the center of gravity is located at the centroid. To find the centroid, you would mark the midpoint of each side of the triangle. Then you would draw a line from each midpoint to the vertex that is across from it (called the opposite vertex), as shown here:

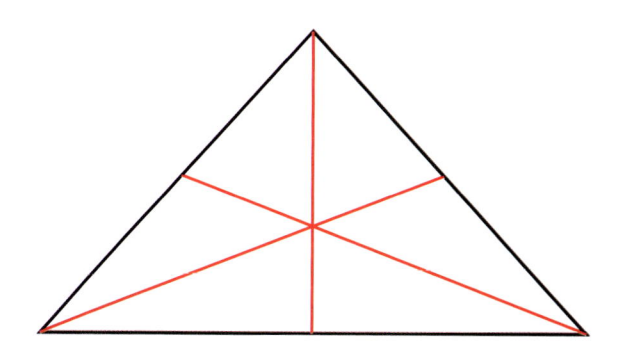

The centroid is the point where the three lines intersect. This is really easy to test out for yourself with any triangle. Cut a triangle out of a sheet of paper and mark the centroid as shown. Then make a tiny hole at the intersection and place the triangle on the tip of your pencil. If you are careful and only make a tiny hole, you should be able to test whether it balances around that centroid or not.

## Tennis is a big racket

Of the games we often watch, one of the ones that might strike you as being inherently geometrical is tennis. Just the mere fact that they are playing on a big court marked off with squares and rectangles of different sizes suggest some of the ways geometry factors into the game.

I think it would be cool to figure out how long it takes a killer serve to travel from the server's racket to the opponent's corner. Besides, this gives us a chance to think about sticking giant triangles all over the court, one of them even standing up vertical like a divider. The diagram for this one is a little complicated, so stick with me as we walk through it.

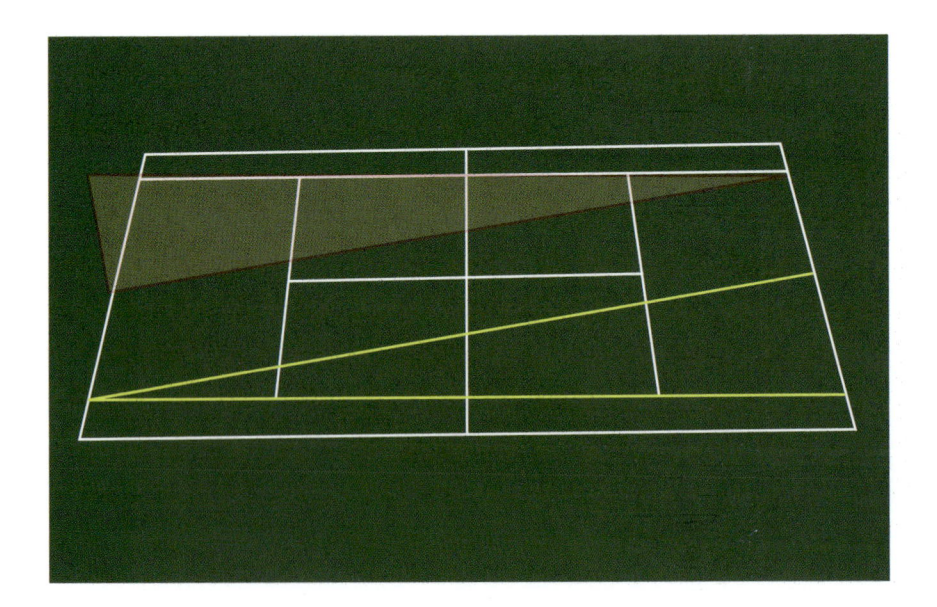

If we look at a common serve, the serving player stands just inside the center line and serves to the feet of the player who is positioned near the corner on her side of the court. This is represented by the gold triangle in the diagram. Notice that given the symmetry of the court, I have put the two triangles under consideration on different sides of the court. That was done just to keep from getting lost when talking about which lengths we're finding.

So, the first thing we need to find is the linear distance along the ground from server to defender. Let's grab ye olde Pythagorean theorem and put it to work. (I know, he wasn't from old-time England. Would you rather I put it in Greek letters?)

$$a^2 + b^2 = c^2$$

The leg along the baseline is 13.5 feet, and the length from baseline to baseline is 78 feet, so …

$$13.5^2 + 78^2 = c^2$$
$$182.25 + 6{,}084 = c^2$$
$$6{,}266.25 = c^2$$
$$79.16 \text{ feet} = c$$

Hang on, because this part is cool. We are now going to take that hypotenuse (that we just calculated) and it will be the base of the pink triangle that is standing upright across the court. The other leg will be the vertical line reaching from the ground up to the height of the tennis player's racket when her arm is extended to hit the serve. If you've watched recently, you'll notice they go for as high a reach as they can. We want to calculate the length of the hypotenuse from the height of a player's serve to the feet of her opponent.

Geometry in Your Everyday Life

The serve of a tennis pro like Serena Williams can be thought of as rocketing along the edge of a triangle.

A little quick research shows me that Serena Williams, one of the best women's tennis players of all time, is 5' 9" (5.75 feet). With overhead extension during the serve and the additional racket length, a good estimate is that she is making contact with the ball at right around 9 feet in the air.

60   Applying Geometry to Everyday Life

$$a^2 + b^2 = c^2$$
$$9^2 + 79.16^2 = c^2$$
$$81 + 6{,}266.31 = c^2$$
$$6{,}347.31 = c^2$$
$$79.67 \text{ feet} = c$$

Now we're in position to figure out how long a player has to react to the powerful serving speed Williams can unleash. Some of her serves have been clocked at 128.6 miles per hour. At that speed, then, let's do the conversions to see how long it takes to travel 79.67 feet.

$$\frac{128.6 \text{ mi}}{1 \text{ hr}} \times \frac{5{,}280 \text{ ft}}{1 \text{ mi}} \times \frac{1 \text{ hr}}{60 \text{ min}} \times \frac{1 \text{ min}}{60 \text{ sec}} = \frac{188.613 \text{ ft}}{1 \text{ sec}}$$

For the last step, we need a proportion:

$$\frac{188.613 \text{ ft}}{1 \text{ sec}} = \frac{78.67 \text{ ft}}{x}$$

Then we do some cross multiplying:

$$1 \times 79.67 = 188.613x$$

Then divide both sides by 188.613:

$$x = 0.4224 \text{ seconds}$$

At that speed, the player receiving the serve has right about the same time to react as a major league baseball batter against a top-level fastball (and the batter doesn't have to hit it to a certain section of the ground closer than third base!).

Geometry in Your Everyday Life

In measuring distances at sea, the theoretical shapes of geometry and the slightly distorted shapes of the real world have to compromise.

62   Applying Geometry to Everyday Life

# Nautical miles

You may have heard the term "nautical miles" before, and wondered why those need to be different than regular old land miles. Originally the term was used to refer to a distance of one minute, but not one minute of time. Rather, it's the distance when each degree of the circle of the Earth is divided into 60 equal distances. You might recognize this in abbreviations of latitude and longitude which are shown as, say, 23°4'15" N (23 degrees, 4 minutes, 15 seconds north).

However, this was based on the idea that the Earth was perfectly spherical. Since the Earth actually bulges a little bit out of shape as you get close to the equator, the results is that those nautical miles were a little longer than the ones at altitudes farther north.

In order to compensate for this, an international unit was established. According to that, a nautical mile is defined to the specific value of 1.852 kilometers (about 1.15 land miles). That means a speed of 60 nautical miles per hour is about the same as 69 miles per hour (111 kilometers per hour) in a car. The nautical speed is usually referred to as "knots." A knot is 1 nautical mile per hour (so you would not say "knots per hour"). This term comes from the way speed used to be measured with a knotted rope.

## Measuring things that aren't even there

If you stop to think about it, one of the fascinating things about geometry is that we use it not just to divide up and measure things—houses, walls, and so on, but we also use it a lot to measure non-things—the emptiness where there aren't any things.

Take the example of a swimming pool. When we talk about the volume of the pool, we're using geometry to figure out how big the empty space in the middle is. So, let's take a few seconds to consider how we would measure the empty space inside a pool before it's filled with water. Let's say we're evaluating an Olympic-size pool, but with a sloped end.

There are different ways we translate this strange shape up into measurable pieces. For instance, we could calculate the **perimeter** around one side of the pool and then multiply that times the width of the pool. I tend to think it's easier and quicker to break it into separate pieces and tackle each piece separately.

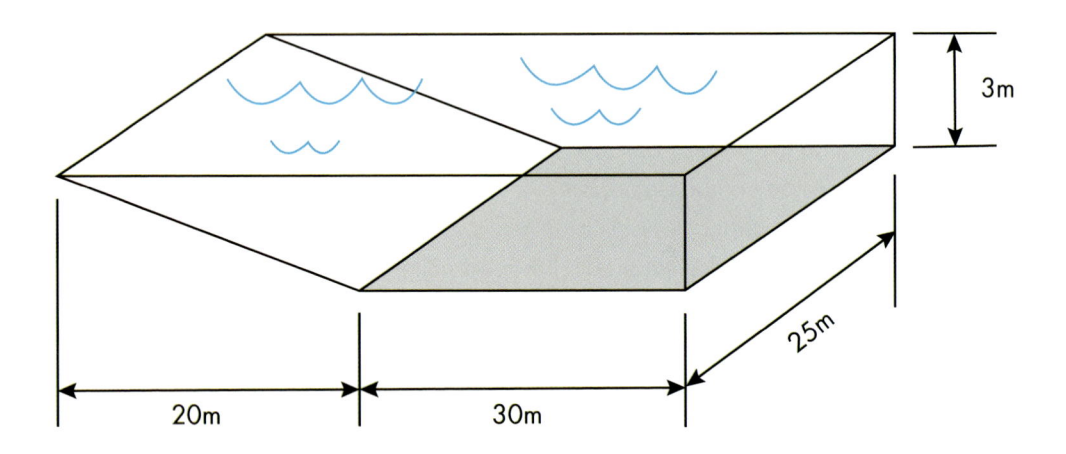

In this case, I would break the problem into a big box that's not there and a triangular **prism**. We imagine this box as if the triangular slope isn't there, and instead the pool is just 3 meters

64    Applying Geometry to Everyday Life

deep everywhere. The dimensions of the big box are 50 meters wide, 25 meters long, and 3 meters deep to allow for diving.

When we run the volume calculation, then, we find:

$$V = lwh$$
$$V = 50 \times 25 \times 3$$
$$V = 3{,}750 \text{ m}^3$$

Now we notice that the box that's not there has a triangle prism that needs to be chopped away. That means, in a way, we need to calculate to take away the part that isn't there from the box that already isn't there. We've got a right triangle, so that means it's time for the Pythagorean theorem? Not this time. I like your enthusiasm, though. No, in this case, we just need the plain old triangle equation:

$$A = \tfrac{1}{2}bh$$

where:

*A* = The area of the triangle.
*b* = The base.
*h* = The length.

$$A = \tfrac{1}{2} \times 20 \times 3$$
$$A = 30 \text{ m}^2$$

Now we need to multiply that times the width of the pool, so:

$$V = \text{Area of base} \times \text{length}$$
$$V = 30 \times 25$$

$$V = 750 \text{ m}^3$$

Then we do our subtraction. This one turns out to be pretty easy:

$$V = 3{,}750 \text{ m}^3 - 750 \text{ m}^3$$
$$V = 3{,}000 \text{ m}^3$$

By using a little simple math, we're able to calculate a complicated volume.

## The Geometry of Movement

In addition to thinking about geometry as something "out there" in the world, it is worth considering for a few minutes that some of our passions and pastimes are about using our bodies themselves to create geometry. This shows up particularly well in the cases of dancers and martial artists.

Not only do the martial arts bring geometry into action through their movements, but different martial arts do so in noticeably different ways. The wide circle of a spinning back kick from an art like tae kwon do is very different from the small circles of the wrists and forearms that are so characteristic of wing chun kung fu. The fast, straight lines of boxing are very different than the clash of intersecting arcs and angles of a **kendo** match.

In very similar ways, we see different styles of dance bring a very different mix of shapes and angles to life. Consider the high, spinning leaps of ballet, or the sharp angles and kicks of modern jazz dance. Think of some of the concepts we've been talking about in other sections—things like balance and symmetry.

As the number of dancers increases, a wider array of possible geometric shapes and patterns opens up. We go from just considerations of body position and the shapes and planes created by movement, but now we add in changes of formation, symmetry and contrast with other dancers, and so on.

Different sports create different shapes in the air. If these kendo swords were sparklers, you could see a circular shape as they moved.

## I'm so board

Maybe you were not expecting something like skateboarding to get lumped in with something like ballet. Consider for a moment, though, how much more like an art form it is than a sport. In order to make either work as a sport, an entire mechanism has been bolted on rating different difficulty levels and so on, but ask yourself: Is the goal of skateboarding to get points the way the goal of throwing a ball at a hoop or shooting a puck is? Skateboarding is not like cycling, where the goal is to get somewhere faster than

In skateboarding and similar sports, the body moves through different geometric patterns.

68   Applying Geometry to Everyday Life

everyone else. It is not like powerlifting where the goal is to hoist heavier slabs of iron.

The goal of skateboarding to get our bodies and boards to go through a series of complicated circles and twists and turns. We are spinning along the axis from head to toe and maybe we are also flipping in mid-air. The board is spinning along a line running from the back truck to the front. It's like a collision of circular planes, really.

OK, show of hands: who saw something surprising in this section? Come on, nobody's watching you, so you can admit that something might have been just a tiny bit cool. Before we go to wander through some different careers and the ways that they end up using geometry to work toward their goals, take a minute to just consider other possible ways that you can picture geometry showing up in your own normal activities.

A few ideas worth holding in mind as you leave the chapter:

- Sometimes geometry concepts don't need numbers.
- Sometimes we are aware of the fact that we are using geometry as part of our reasoning process, but other times it is below the surface.
- Artists have just as much claim to being geometers as do geometry teachers and hang glider designers.
- Skaters have just as much claim to being artists as dancers do.

✢ In some traditional Zen gardens, patterns of raked gravel stay still but imitate movement, like circular ripples in a pond.

# THREE

# Geometry in Others' Everyday Lives

Now that we've reflected on the volume of things and emptiness that are parts of our day-to-day lives, let's turn to consider the role geometry plays in the lives and careers of those whose jobs touch on the subject. We'll even spend a few brief minutes thinking about a **Zen** rock garden, which is kind of the opposite of work.

Sometimes our models and processes for thinking geometrically about things don't always have numbers or calculations involved. Because of that, people for whom geometry is important often don't think of that as "having math" in their jobs. A great example of this was when I spoke with a physical therapist and just seconds after saying she didn't think there was really all that much math in her everyday work life, she told me how when she looks at clients she is working with, she is imagining where the axes of rotation go through the body. She also mentioned how it can seem a little bit like looking at people who have little circles taped all over them because in her head she is always walking through the process of watching for subtle cues that this or that plane of rotation might be tilted out of alignment or limited in some way.

When I pointed out that this was a very mathematical way of looking at the world around us, she laughed. I hope you will reflect on that idea as you head off into this next chapter. Sometimes geometry is about trying to capture the whole universe in a single picture, like with Tibetan **mandalas**. I hope that you, too, are able to notice some things that you do but just didn't think of as math. I also hope that you find lots of things that shape people's views of the world. If nothing else, at lease be a little impressed when we talk about the Zen gardens. Those guys take thinking about emptiness to a whole other level.

## Archaeologist

Archaeologists (and anthropologists, a closely related field) study the evidence, remains, and artifacts of historical groups in order to help us understand their cultures, social structures, and ways of life.

To accomplish this, they combine a life of academic research looking through historical documents and records, scientific developments, and the work of other researchers with hands-on work in the field. The job ranges from digging in the soil of historic sites to crunching numbers on computers to look for variations and patterns in data about a culture or a group of people under investigation.

Archaeologists also play an important role in helping to preserve these sites and artifacts. They work to educate others about what the past has to teach us regarding the shape and possible directions of our own cultures and ways of life.

### Let's play a little basket bowl

The basket on page 73 was woven by the Pima Indians, a tribe native to modern-day New Mexico. It has a kind of symmetry known as "rotational." How many degrees around is each successive rotation in this pattern?

Well, we know that traveling full circle is 360 degrees. In this case, the basket shown has seven repetitions of the L-shaped pattern. So when we take the general equation:

$$x = 360/n$$

where *n* equals the number of repetitions, we know each L shape occupies:

$$x = 360/7 = 51.43 \text{ degrees}$$

That's an easy enough puzzle to solve when looking at an example of the entire basket. Often the archaeologist is working from fragments of dishes or fabrics and trying to work backward to reconstruct what an intact original would have looked like. By understanding how these kinds of equations work, she can reverse the process.

## Mandala

This painting from a temple ceiling in **Tibet** is remarkable. In trying to understand the production and use of specially colored paints, the archaeologist would use geometry to figure out an estimate of how much area is covered with that deep blue paint.

Ancient mandalas like this one use geometric symbolism to represent the complexity of the universe and the afterlife.

Because estimation tools are more useful on big objects than small, she begins just with a measurement of the central circle, placing it at about 10 feet in diameter. If we start there, how would we go about forming a fair estimate for the area covered in blue?

If the central circle is 10 feet in diameter, then it is close enough for estimate purposes to have the length of the central square be 10 feet as well.

Our first step is to calculate how much blue there would be if the central square were filled solid, and then subtract out the area enclosed within the central circle.

$$A = lw$$

where:

$A$ = Area of square or rectangle.
$l$ = Length of the shape.
$w$ = Width of the shape.

$$A = 10 \times 10$$
$$A = 100 \text{ ft}^2$$

For the central circle, our general formula is:

$$A = \pi r^2$$

The radius will be half of the diameter, so 5 feet.

$$A = \pi \times 5^2$$
$$A = \pi \times 25$$
$$A = 78.54 \text{ ft}^2$$

Geometry in Others' Everyday Lives

The result is 100 – 78.54, which gives us 21.46 square feet.

There are twelve smaller squares which surround the larger square. So we must calculate the amount of blue used for one of the small squares, and then multiply by twelve and add to the blue in the central square. We need not actually walk through the whole process, though you are free to take it and run with it on your own. The key ratio we need to use in order to proceed is provided by noting that each of the smaller outside squares is one half the height of the inner square. From there, you can find both the area of the squares and the area of the circles inside to subtract away.

Sometimes, though, our **intrepid** archaeologist is not so lucky as to have photographs, but might, instead, be working from notes and sketches—either her own or those from other, earlier archaeologists. In calculating the area of blue that would have been used in the temple ceiling mandala shown below, what would be a reasonable estimate?

The only dimensional notation given in the tattered old field journal the archaeologist found said that the outside border seemed to be about 12 feet by 12 feet. This was from a smaller temple than the one depicted above, so it might have been harder to seek it out for photographs—if the temple still existed at all.

So, let's begin by finding the area of the outside square, and work in from there.

$$A = lw$$
$$A = 12 \times 12$$
$$A = 144 \text{ ft}^2$$

Next, we need to subtract out the entire area contained within the largest gold circle, so we need to figure out the dimensions of that circle. The easiest way to do that, since we're just working for an estimate, is to pretend that it actually reaches a tiny bit farther to touch the edges of the outside square. That allows us to see that its diameter is the same as the square—12 feet. For the area of a circle, we know that:

$$A = \pi r^2$$
$$A = \pi \times 6^2$$
$$A = \pi \times 36 = 113.097 \text{ ft}^2$$

That brings the blue area down to 144 − 113.1 = 30.9 ft².

Next we need to calculate the area of the outermost four white circles in the corners. We can find that by first subtracting the radius of the larger circle from half of the diagonal of the largest square. That will give us the diameter of one circle.

Geometry in Others' Everyday Lives

For the diagonal of the square, we're going to have to rely on our old friend Pythagoras and his theorem:

$$a^2 + b^2 = c^2$$

where:

*a* and *b* = The lengths of the legs of a right triangle.
*c* = The length of the hypotenuse.

$$12^2 + 12^2 = c^2$$
$$144 + 144 = c^2$$
$$288 = c^2$$

Now we take the square root of both sides and get:

$$16.97 \text{ ft} = c$$

Now since we only want to compare to the radius of the circle, we divide that result by 2 to get 8.485 feet. We'll subtract half of the square's length (the radius of the large inner circle):

$$8.485 - 6 = 2.485 \text{ feet}$$

We'll round that to 2.5 feet to keep our fingers from falling off keeping track of all those decimal places. Now we know that the diameter of one of the outermost corner circles is 2.5 feet, which gives a radius of 1.25 feet. Going back to our formula for the area of a circle, we can plug in the value:

$$A = \pi \times 1.25^2$$
$$A = \pi \times 1.5625$$

# Space archaeology

When we picture archaeology, most of us imagine the old-fashioned process of digging and sifting to find buried fragments of pottery to show the edge of an old settlement deep in the jungle or a few well-preserved bones to mark some ancient burial site. Admittedly some of us still imagine there's an occasional Indiana Jones moment mixed in as well. What we probably don't imagine is high-tech satellites beaming rays into the ground to search for important structures under the surface.

Archaeologists and anthropologists from different universities are starting to **leverage** this in their search for sites and cities around the world. The Egyptian capital city during the Middle Kingdom, temples in ancient Cambodia, and trade routes in Persia are just a few examples of using this technology.

This process involves beaming lasers at the ground—about a million of them every five seconds—and then measuring how long it takes them to bounce back to the satellite. (No, they're not *that* kind of laser. They're very small and weak and just used to measure stuff. No planets were harmed in the making of this research.) This allows the scientists to detect small changes in **elevation** (possible buildings underneath) or soil density (as would be expected of a highly-traveled trade route) that would otherwise be undetectable. This strategy can save years of time archaeologists would spend searching on the ground.

$$A = 4.90874 \text{ ft}^2$$

Four of those would total up to 19.635 ft². If we grab the earlier total for blue, we can update it:

$$30.9 \text{ ft}^2 - 19.635 \text{ ft}^2 = 11.265 \text{ ft}^2$$

But the calculation process isn't done yet. To get the final answer, our archaeologist would have to go on to calculate the information for the blue area inside the smaller square. Because of the self-similar nature of many mandalas, such as this one, the process would proceed along identical steps as the first one did, and at the end, she will add 11.265 ft² with her new calculation to get a total blue area value.

## Release the mounds

One of the largest ancient cities was in the middle of the United States, near Collinsville, Illinois, which is just a little ways east of St. Louis, Missouri. Around the year 1250 CE, the city was bigger than London. The Native Americans of this culture built a series of more than 120 giant, earthen pyramids. They range in size, with the largest one being Monk's Mound.

Monk's Mound has a rectangular shape with a base measuring approximately 950 feet (290 m) long and 836 feet (255 m) wide. It is a little over 100 feet (30 m) tall (as tall as a ten-story building). In order to calculate the volume of one of these pyramids, our archaeologist has to do a little bit of prep work first. By setting poles atop the flat-topped structure, she can imagine the edges continuing up to see where they would have met if it had been a complete pyramid.

It is hard to tell because, unlike stone pyramids, these earthen mounds were subjected to much weathering over time. But if the faces of the longer sides of the pyramid have an estimated slope of 45 degrees, we can calculate a theoretical height.

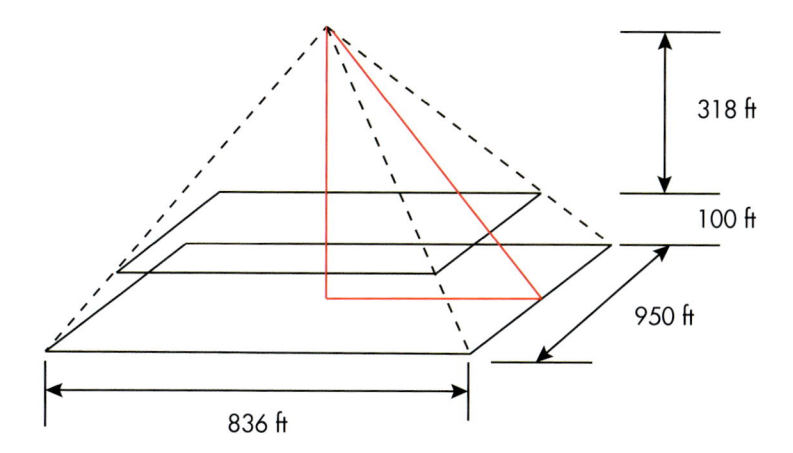

The archaeologist can get a quick estimation by using a little geometry. If she runs the calculations on the red triangle, she finds that the length of the base would be the same as the height, which is a property of 45-45-90 right triangles:

$$½ \times 836 = 418 \text{ ft}$$

Now, to find the volume of that completed pyramid, we need to calculate:

$$V = \tfrac{1}{3} lwh$$

where:

$V$ = Volume.
$l$ = Length of the base of the pyramid.
$w$ = Width of the base.
$h$ = Height of the pyramid.

$$V = \tfrac{1}{3} \times 950 \times 836 \times 418$$
$$V = 110{,}658{,}533 \text{ ft}^3$$

Geometry in Others' Everyday Lives

Now the people who built this structure never built it anywhere near that height, remember. This is just theoretical. To find the volume of the truncated pyramid shape they actually did build, we now need to subtract away the volume of the imaginary top part of the pyramid.

The base of that pyramid would be 636 feet wide. Going up 100 feet means we'd lose 100 feet of width on one side, given the 45 degree angle, and that would happen on both sides. The proportional change in the other two sides, which don't have the exact same slope, would be:

$$636/836 = x/950$$

$$636 \times 950 = 836x$$

$$604{,}200/836 = x$$

$$722.73 \text{ ft} = x$$

Now we use those to calculate the volume of the top pyramid:

$$V = \tfrac{1}{3} lwh$$

$$V = \tfrac{1}{3} \times 722.73 \times 636 \times 318$$

$$V = 48{,}723{,}566 \text{ ft}^3$$

Once we put those two numbers together, 110,658,533 ft$^3$ − 48,723,565 ft$^3$, we are left with a trapezoidal pyramid structure containing about 62 million cubic feet of soil. Is that not a number that relates for you? How about this: that's almost enough dirt to fill up a modern football stadium. So, yeah, you can kind of imagine it if you filled up a giant stadium with dirt by hand and then turned it over like making a sandcastle!

# Interior Designer

Interior designers work to **optimize** the usefulness and appeal of various indoor spaces. They work with almost every aspect of a building's interior from floor covering to lighting placement and from windows to wall color.

Their jobs include meeting with clients to discuss goals and hopes for a given space, consulting with experts on lighting fixture options, and meeting with furniture designers and architects. They then take those inputs and work with a computer program to come up with recommendations for helping to achieve the goals the clients have identified.

The interior designer would also have the role of project manager. That job includes creating and managing the timelines for different threads of a creation or remodeling project. The designer is also the one who oversees the order of materials and needed equipment. They would continue to meet with the client periodically throughout the project for feedback and **collaborative** decision making.

## You floored me

Let's start with a simple example of how geometry could come into play in the course of an interior design project. Naturally there will be such predictable problems as figuring out how much paint is needed to cover $x$ amount of living room wall and checking to see if furniture of these dimensions will possibly fit through that door. Let's try something a little less easily undone or covered up.

Suppose our trusted interior designer is trying to hang a curtain rod, making sure it is parallel to the ceiling. Now you might think, "Hey, I'll just use my super-amazing tape measure." But the trick is, there is no easy way to make sure you are holding your tape measure exactly parallel to the ceiling—a plumb line only guarantees that it will be **perpendicular** to the earth, but most interior designers will warn you that this depends on the faulty assumption that the house still remains as perfectly perpendicular as when it was born.

Instead, he can use a piece of string and a pencil as if it were a great big compass (the kind for drawing circles, not the kind for navigating deep wilderness). He measures the string and marks it at the desired distance from the ceiling. Then, holding one end at the edge of where the wall and ceiling meet, he draws a small arc in the area of where he wants to hang the rod. Then he'll move over a couple feet and repeat the process. He'll hang one end of the rod where the two arcs cross. Then, after measuring the length of the rod, he'll repeat the process of making two arcs at the other end.

Notice that the diagram shows a badly tilted ceiling. Yet the rod will hang parallel to it.

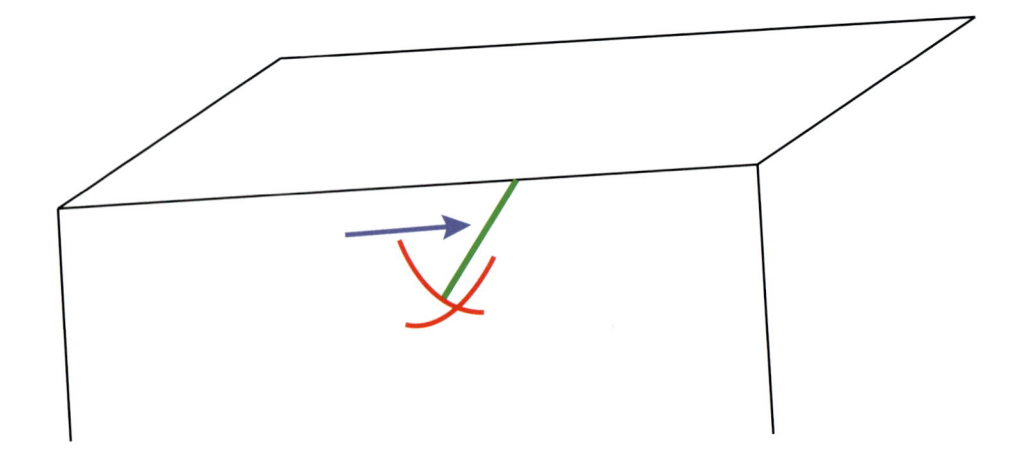

## A room with a view

As part of the remodeling process, our homeowner wants to convert part of the house to a traditional Japanese-style room to use for meditation and for special occasions. In traditional Japanese houses, the interior walls were not load bearing, which allowed them to be made of movable partitions. Because they used special mats, called tatami, made of rice straw and woven reeds,

> Traditional Japanese house designs allowed several possible combinations of differently sized rooms.

they would often design their houses and arrange their rooms in terms of these mats. They might have, for instance, a six-tatami room, or a ten-tatami room. These mats are traditionally 3 feet by 6 feet in size.

The partitions cannot go over a tatami, but can go between any long breaks between them. Since any length or width of a room we create would have to be divisible by a factor of 3 feet, what are the possible room dimensions for a ten-tatami room?

Well, first we need to figure out how much total area we are working with.

$$A = lw$$

To help keep us from getting confused when talking about the different areas, we'll use:

$A_t$ = Area of a tatami mat.
$A_r$ = Area of a room.

$$A_t = 3 \times 6 = 18 \text{ ft}^2$$

Since there are ten mats:

$$A_r = 10 \times 18 \text{ ft}^2$$
$$A_r = 180 \text{ ft}^2$$

Now we just need to consider all the factors of 180 and look at pairs that are multiples of 3:

- 3 × 60: OK, but a room that shape would be a hall, and pretty hard to use for other stuff!
- 6 × 30: Barely better.
- 9 × 20: Nope, 20 is not a multiple of 3.
- 12 × 15: Hey, this could work. Test out with some dominoes or scraps of paper to see if you could actually arrange ten tatami in this way with none sticking out.
- 15 × 12: We just did this one, but I wanted you to see how we are following a pattern so we know our solution has been systematic, and therefore less likely to make a sloppy oversight mistake.
- 18 × 10: Nope, 10 is not a multiple of 3.
- 21 × ?: 21 doesn't go into 180 an even number of times.

So a room size of 12 feet by 15 feet (or vice-versa) is the best size if we have ten tatami.

# Farmer

A farmer is engaged in the tending and harvesting of any variety or combination of crops and livestock. Her job varies significantly depending on what she raises, but there are some aspects that most farmers would have in common, including the maintenance and operation of large equipment, and the management and upkeep of the property itself (weed control, fencing, mowing as needed, etc.). Farmers also need to repair and care for their house and the various outbuildings, storage containers, and so on.

If a farmer raises livestock, she will have additional ongoing concerns involving medication, feed content and rates, and decisions about what proportion of the herd is kept, bred, or sold. If, on the other hand, she raises crops, there will be much greater emphasis on issues like irrigation, pesticide applications, storage, and the additional large equipment needed for planting and harvesting.

## Silo maintenance

Because some of the grains that are stored in silos create a kind of acid, a farmer needs to coat the inside of the silo with a special compound from time to time. If one gallon of compound can cover 225 square feet, how many gallons would she need to buy to resurface a silo that is 54 feet from the base to the top of the dome and has a diameter of 20 feet?

First we need the surface area of a sphere:

$$A = 4\pi r^2$$

We'll cut that in half because we have a dome, not a whole sphere:

$$A_d = 2\pi r^2$$

With only a little imagination, we can easily picture how the different parts of a farmstead are comprised of prisms, cylinders, and boxes.

where:

$A_d$ = The surface area of the dome.
$r$ = The radius of the dome.

$$A_d = 2\pi \times 10^2$$
$$A_d = 2\pi \times 100$$
$$A_d = 200\pi = 628.32 \text{ ft}^2$$

Now we need to calculate the surface area of the cylinder section of the silo. Since we know the total height, we can subtract 10 feet for the dome, giving us a height of 44 feet. The general formula for the surface area of a cylinder is:

$$A = 2\pi rh + 2\pi r^2$$

We only need half of the last part, since that is to account for both the top and bottom, but in this case the top of the cylinder is the dome. We've already calculated the surface area for that.

$$A = 2\pi rh + \pi r^2$$

Now we plug in our height and radius:

$$A = (2\pi \times 10 \times 44) + (\pi \times 10^2)$$

Then wave a magic calculator over the numbers:

$$A = 880\pi + 100\pi$$
$$A = 980\pi = 3{,}078.76 \text{ ft}^2$$

To find the total surface area, we add that to the dome:

$$A_s = 3{,}078 + 628.32 = 3{,}707.08 \text{ ft}^2$$

Given that one gallon of compound covers 225 square feet, the farmer needs to buy 3,707/225 = 16.48 gallons. Since it's impossible to find containers with exactly 0.48 gallons, the farmer will buy 17 gallons.

## Adding a corral

Our farmer has 200 feet of spare fencing and she wants to use it to put a rectangular corral on the back side of her barn, using the barn as one side of the pen. What are the dimensions for the largest possible area she could construct?

In this case, we're starting with two equations. We already know the area equation:

Geometry in Others' Everyday Lives

$$A = lw$$

Since the perimeter adds up all sides of a rectangle, and the barn is one of those sides, we only need three here. Remember, the 200 is for the total feet of fencing available.

$$2w + l = 200$$

We'll get *l* by itself to set up a substitution:

$$l = 200 - 2w$$

Then we can substitute for *l* in the area equation:

$$A = (200 - 2w) \times w$$

Distributing gives us:

$$A = -2w^2 + 200w$$

This is in the form of a quadratic equation, and the graph would be a parabola. We can imagine mapping it onto *x* and *y* coordinates to quickly find where it would reach the highest area value. We know that the *x* coordinate for the maximum value of a quadratic equation like this is at:

$$x = -b/2a$$

The *x* coordinate for the vertex is:

$$x = -200/(2 \times -2)$$
$$x = 50$$

Now we plug that back into our area equation. Note that in this case, area (*A*) represents the y-axis.

$$A = -2 \times 50^2 + 200 \times 50$$
$$A = -5{,}000 + 10{,}000$$
$$A = 5{,}000 \text{ ft}^2$$

The only thing left is to find the *y* coordinate value (which stands for the length):

$$5{,}000 = 50y$$
$$100 = y$$

So the dimensions of the corral should be two sides of 50 feet, and the long side, parallel to the barn, should be 100 feet.

When you're putting up fence posts for the corners, though, it can be tricky to make sure you're getting them just square. Think about it: Can you picture a farmer out there trying to measure the angle of the intersection with a little protractor?

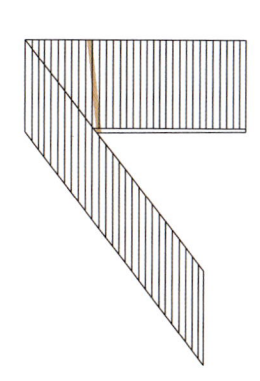

For this, you use a little trick. Get yourself a piece of bailing twine or rope and knot it off at 5 feet (or five sections of the fence, depending on the material of fencing). You tie it 4 feet from the end of one piece of fence. Then you tie it 3 feet from the end of the other piece of fence that is meeting the first end to make a corner. Put the ends of the two fence sections together and open them until the twine hypotenuse is drawn taut.

Now you've just used the Pythagorean theorem in much the same way ancient Egyptians would have used it to re-establish

# Meet Jean van Booven-Shook

Please meet Jean van Booven-Shook, owner and lead project manager for All Flooring Kansas City.

*So, tell us: How long have you been doing flooring?*

Since the summer of 2012. I worked for someone else for a few years before going out on my own.

*What are some different ways and kinds of math that you use?*

At this point, math is nearly everything I do, and I could not survive if I didn't know how to use spreadsheets and program the formulas I need. Basically, when someone needs a flooring project done, they either send me drawings and a description of what they want done by email, or they ask me to come out in person to look at what they want done.

Geometry is a big part of what I do. For instance, one time I talked with a customer about why it made sense to have our company recarpet their community center. One of the rooms in the community center was a circle. We had to figure out how much carpet we needed, but also how much wall base.

Another time, a customer's store had main walls that ran parallel, but one was longer than the other. We had to figure out how much square tile we needed for the triangular part.

In real life, there are other factors to consider. Often, customers want two different kinds of flooring in different parts of the building, and each kind of flooring requires a different kind of preparation and adhesive. Another thing to keep in mind is that industry talk uses different standards of measure to talk about different materials. For example, ceramic tile is typically quoted per foot, where carpet is quoted per yard.

*Why do you have to use good geometry? What goes wrong if your math is off a little bit here and there?*

If my math is wrong, I might end up buying too much product and having some left over. There usually isn't enough to sell or do another project with. What is worse than buying too much product is buying too little. Most of our carpets and tiles have to be ordered ahead of when we plan to install. We don't want to have someone's office or store torn up and find out we don't have enough carpet or tile to cover the floor, and that it will take a couple weeks to get more!

field boundaries after the Nile flooded each year. Doing this same trick with one of the legs as the side of the barn will help you make sure you set the two sides perpendicular to the barn. By doing that, you know the long section of the fence will be parallel to the barn, as well.

## So, water you doing?

As our farmer thinks about keeping some of the livestock in the corral, she realizes she will need to upgrade the water tank because the one she's been using won't be big enough to prevent crowding.

If she's been using a cylindrical water tank with a radius of 3 feet, what radius tank would she need to have delivered in order to have twice as much water at one time and twice as much room for livestock to line up around it? Logic suggests that double the radius should be right. Let's test it out.

The general formula for the circumference of a circle is:

$$C = 2\pi r$$

where:

$C$ = The circumference.
$r$ = The radius.

So the current tank would have circumference of:

$$C = 2\pi \times 3$$
$$C = 6\pi$$
$$C = 18.85 \text{ ft}$$

If she switches to a radius of 6 feet, then the result would be:

$$C = 2π \times 6$$
$$C = 12π$$
$$C = 37.70 \text{ ft}$$

So doubling the radius will get us twice as much room. Does it accomplish her goal of also having twice as much water? For that we have to calculate volume, which works from the formula:

$$V = πr^2h$$

where:

$V$ = Volume of a cylinder.
$r$ = The radius.
$h$ = Height.

Her current tank has a height of 2 feet and a volume of:

$$V = π \times 3^2 \times 2$$
$$V = 18π$$
$$V = 56 \text{ ft}^3$$

One cubic foot is 7.48052 gallons, so that translates to 56 × 7.48 = 418.9 gallons.

If we doubled the radius, we'd have 6 feet, also with a height of 2 feet. Its volume is:

$$V = π \times 6^2 \times 2$$
$$V = 72π$$
$$V = 226.19 \text{ ft}^3 = 1{,}692 \text{ gallons}$$

That's four times as much water! It would take forever to fill that tank. How could our farmer solve the puzzle of getting way more room but not having to waste so much water? Think. With geometry, one of our strategies is always to consider chopping our shapes in half. No, that doesn't mean literally cutting the watering tank in half, but it would work if she just filled it half full, wouldn't it? That would mean only two times as much water instead of four times.

## Hay, what's up?

When Farmer Applegate runs the round hay baler, it puts out bales that are 5 feet long and have a diameter of 5 feet. If he mows about 6,000 cubic feet of hay, how many bales will that produce?

Well, let's figure out how many cubic feet of hay end up in one bale. The formula for cylinder volume is:

$$V = \pi r^2 h$$

where:

$r$ = The radius of the circle.
$h$ = The height—or in this case the length.

$$V = \pi \times 2.5^2 \times 5$$
$$V = \pi \times 6.25 \times 5$$
$$V = \pi \times 31.25$$
$$V = 98.175 \text{ ft}^3 \approx 98.2 \text{ ft}^3$$

Now we can calculate:

$$6,000/98.2 = 61.099 \approx 61 \text{ bales}$$

Those bales are going to come in around 1,100 pounds each. Working from a rule of thumb, cows are going to eat about 32.5 pounds of hay per day, so this gives him enough hay to feed about seventeen cows through four months of winter.

## Barn to be wild

That amount of hay won't be quite enough, so the farmer wants to be able to buy some square bales to store in his barn loft. How much hay would fit if the loft has the following dimensions:

The formula for any prism is the area of the base times the length (or height, depending on orientation). For the loft of the barn, we will break it into a triangle and a trapezoid. Then we can add those together and multiply by the length of the barn.

To find the area of the trapezoid, we need the bottom width, which is 24 feet, and the width of the top yellow line, which is 18 feet. We also need the height, which is 10 feet. We can take the general formula for the area of a trapezoid:

$$A = \tfrac{1}{2}(b_1 + b_2)h$$

# Meet Richard Lierheimer

Please meet Richard Lierheimer, a lifelong farmer with a bachelor's degree in agricultural economics.

*Thank you for sharing your thoughts with us, Richard. Can you tell us a little bit about how math plays into farming?*

Well, you know, it comes in from several different directions because you have the farming itself, and you also have the fact that a farmer is always running a small business, too. That means you have things like trying to understand trends in prices, and on the day-to-day level you have all the things that help you raise good, productive crops.

*Interesting. Are there ways that geometry comes into it?*

A lot of times we don't even think about the fact that we're using geometry. It has become a habit and just seems like the common sense way to do things. Take harvesting, for instance. With modern planting, it works the best to plant going at one angle and then you run the combine at a different angle, so you're cutting across the rows. In order not to lose grain when you're turning around at the end of the rows with each pass, you have to start the field by making one or two circuits around the perimeter. That clears out enough area for you to make the turns.

*Are there places where you do have to work through some of the calculations for stuff like area or volume or any of that?*

98   Applying Geometry to Everyday Life

Sure. Take our irrigators, for instance. Each of those automated irrigators stretches out from a center hub, so they make a circle as they pass, even if it's a square field. You have to take into account how fast they travel, how much area they are covering, and then figure out the amount of water to pump through to get the amount of "rainfall" you need—maybe that's half an inch, maybe that's six-tenths.

You're also working all the time with estimates of how much yield you expect so you can estimate the profit or loss when you invest in seed. On modern equipment, a lot of those functions are computerized, figuring out your harvest and so on, but if you don't know the nuts and bolts of those calculations, you don't know what to really do with the numbers it's giving you.

*That's great, Richard. What a surprise for most of the readers this will probably be!*

where:

$b_1$ = Length of base 1, the bottom yellow line.
$b_2$ = Length of base 2, the top yellow line.
$h$ = The height of the trapezoid.

$$A = \tfrac{1}{2}(b_1 + b_2)h$$
$$A = \tfrac{1}{2} \times (24 + 18) \times 10$$
$$A = \tfrac{1}{2} \times 42 \times 10$$
$$A = 21 \times 10 = 210 \text{ ft}^2$$

Then we need to find the area of the top triangle. There we see that the height is 13 − 10 = 3 feet, and we know that the base is the same as $b_2$ of the trapezoid. That gives us enough information to plug into the formula for the area of a triangle:

$$A = \tfrac{1}{2}bh$$
$$A = \tfrac{1}{2} \times 18 \times 3$$
$$A = 9 \times 3 = 27 \text{ ft}^2$$

Adding those two together, we get 210 + 27 = 237 square feet. Then we just need to multiply that area times the length of the barn, which is 75 feet, and we will have the loft volume.

$$V = A_b l$$
$$V = 237 \text{ ft}^2 \times 75 \text{ ft}$$
$$V = 17{,}775 \text{ ft}^3$$

Since the volume of a standard square hay bale is 5.625 cubic feet (3 feet × 18 inches × 15 inches), it looks like the farmer will have room for 3,160 bales. While the weight can vary depending on the

100   Applying Geometry to Everyday Life

setting on the hay baler, the moisture of the hay when it's baled, and the kind of hay, a good working measure is about 60 pounds per bale. That would mean she has the potential to store 189,600 pounds of hay. Remember that a cow can eat 32.5 pounds of hay per day. The farmer would need to be sure to match the weight of the hay he's stored with the number of cows who need to eat it.

## Landscape Architect

Landscape architects are responsible for the design and creation of the various outdoor spaces (and occasionally indoor spaces that emulate outdoor spaces). This ranges from placement and management of parking and access routes to the selection and placement of trees, ground cover, decorative plants and flowers, and other features involving the use and enjoyment of the outdoor space.

To achieve this, they use a number of different resources and wear a variety of hats. Some of the key tools of their trade include computer programs for landscape planning and design, physical scale models, and site and environmental area studies. Their different roles include meeting with clients and architects to coordinate the different aspects of the project, talking with public government and civic groups, and consulting with technical specialists, like environmental scientists, botanists, and engineers.

### Sprinkler system

Dr. Pockley is having the landscaper install an automated sprinkler system in the front yard of his estate. The front yard is a rectangle approximately 90 feet wide and almost 120 feet long. If each sprinkler covers a radius of 9 feet, how many sprinklers will need to be installed so that all of the lawn will be watered?

It might seem like we need to start by calculating some areas for this one, but we actually don't. All we really need to do is think about the diameter of the sprinkler circles and compare that to the dimensions of the yard.

Since each sprinkler covers a diameter of 18 feet, we just need to think about how many 18-foot circles would fit into a rectangle with the dimensions of the yard. From side to side, we have 90 feet, and that divides by 18 exactly (5). The length of 120 feet does not evenly divide by 18, but comes out to 6.67. Rather than create additional gaps in the sprinkler coverage, Dr. Pockley asks them to round up to 7. Thus, the total number of sprinklers installed for the front lawn would be 35.

102   Applying Geometry to Everyday Life

However, as you can see from the diagram, this leaves some gaps between the edges of the circles. The designer could compensate for that by slightly overlapping the circles, but then that causes grass in those areas to get over-watered. Rather, it makes sense instead to adjust the amount of time the sprinklers stay on in order to make sure that the water going into the soil is able to spread into the areas not being directly sprinkled. For that, we need to calculate the percentage of uncovered grass area.

First, we need the total area of the yard:

$$A = lw$$
$$A = 120 \times 90$$
$$A = 10{,}800 \text{ ft}^2$$

Now we do need to find that area covered by each of the sprinklers:

$$A_s = \pi r^2$$
$$A_s = \pi \times 9^2$$
$$A_s = \pi \times 81 = 254.47 \text{ ft}^2$$

Since there are 35 sprinklers, we simply multiply:

$$254.47 \times 35 = 8{,}906.45 \text{ ft}^2$$

That means the total area not covered is 10,800 − 8,906.45 = 1,893.55 square feet.

If we divide that by the total area, that means 17.5 percent of the yard not directly sprinklerized. (Yes, I just made that up.)

The landscaper explains that this would be the highest coverage but would involve a lot of digging and installation. He offers an alternative that uses fewer sprinklers with a 15-foot radius instead.

How would the coverage look different in that case? You should be able to take a stab at it following the same steps we just walked through. Make a hypothesis about whether you think the overall coverage percentage would go up or down.

The problem in real life would be complicated just slightly by the fact that there is a sidewalk through part of the yard which would displace some of the sprinklers, but this is still the right starting point, even if we were going to go on to make those adjustments.

## Zen rock garden

Out in the back of his house, with a view from the Japanese-style room we discussed, Dr. Pockley is having a Zen rock garden put in. Because of the way the properties in his neighborhood are laid out, his back yard is actually slightly smaller than his front yard, with total dimensions of 90 feet wide and 70 feet deep. In order to leave some room to one side for a stone pathway to the garage, he is going to use only 45 feet of the width.

Dr. Pockley already has a stone and concrete wall around the outside edge of his property, but he would like to add an inside wall that closes off the rock garden, allowing access only from inside the house. Give the way the area is laid out, calculate the total volume of concrete for a wall along the two sides of the garden that is 5 feet high and 1 foot thick. Then figure out how many cubic feet of gravel will need to be brought in if the rock garden needs a layer 4 inches thick to allow for raking patterns.

First, let's calculate for the two new sections of wall. The one in back is 70 feet long and the one at the side of the house is 25 feet long, so we have a total of 95 feet of length. The general formula for volume for rectangular prisms like these is:

$$V = lwh$$

# Zen rock gardens

The tradition of the Zen rock garden goes back to the Muromachi period in Japan, which lasted roughly from the mid-1300s to the mid-1500s. The simplicity of these gardens is balanced by their use of curves raked into the gravel and by the individual stones that stand in contrast to the generic sameness of the large expanse of gravel. Sometimes this contrast is accentuated with the inclusion of some bit of greenery in the midst of the garden or around the edges.

There had been rock gardens of a different form for several centuries prior, but it was not until the fourteenth century that the spread of Zen gave rise to the new form of gardens. These began in the temples of Kyoto, the former capital city of Japan, and gradually spread outward, like ripples in, well, gravel. The first is credited to Musō Kokushi, who transformed an existing temple into Saihō-ji Zen Temple and designed the first garden of the new kind.

At the high point, there are estimated to have been thousands of Zen rock gardens at temples throughout Japan, as well as at the houses of the feudal samurai class. While many of those temples and gardens have disappeared over time, with the spread of Zen to other countries, the Zen rock garden has gone with it.

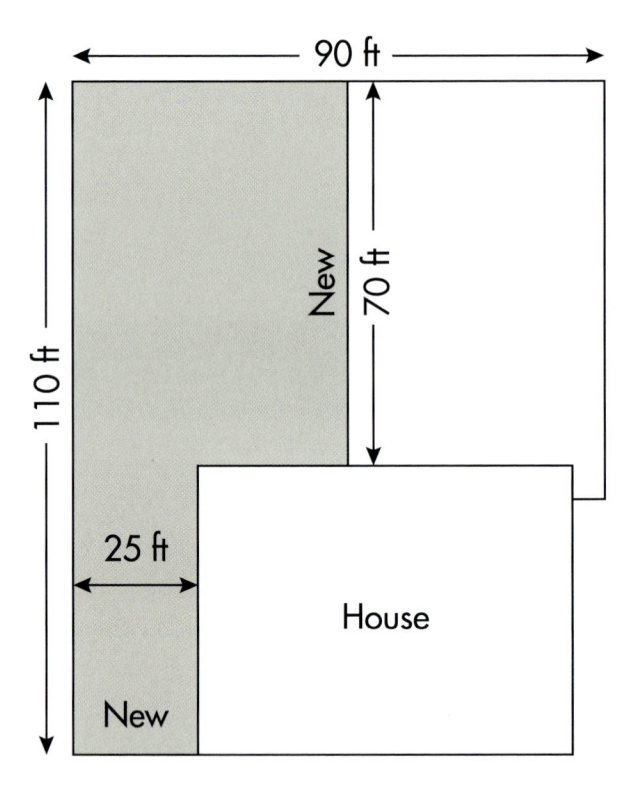

So we can plug in the values:

$$V = 95 \times 1 \times 5$$

$$V = 475 \text{ cubic feet of concrete}$$

Now we can go for a calculation of the gravel that needs to be brought in. As with the front yard, there will be some slight adjustments to the overall total because there will be some larger rocks placed in different locations in the Zen garden, so those areas won't need gravel. No need to worry, it will just be spread in with the rest of the gravel in the other areas.

To find the volume of gravel needed in this case, the logical strategy is to break the space into two rectangles and then add

106  Applying Geometry to Everyday Life

them together to get a total area, and then we will multiply that by the depth of gravel. Let's do the long rectangle first:

$$A_1 = lw$$
$$A_1 = 110 \times 25$$
$$A_1 = 2{,}750 \text{ ft}^2$$

This covers the narrow part extending from beside the house all the way back. The other rectangle is 70 feet long. It can only be 45 − 25 = 20 feet wide, because Dr. Pockley is only using half of his yard for the garden.

$$A_2 = 70 \times 20$$
$$A_2 = 1{,}400 \text{ ft}^2$$

When we add them together, we get a total area of 4,150 square feet. Now, to find the volume, we need to multiply that surface area times the height (or, in this case, depth).

$$V = lwh$$

We'll plug in our number and a value for 4 inches (1/3 feet). Remember we've already calculated $l \times w$ as our total area.

$$V = 4{,}150 \times 1/3$$
$$V = 4{,}150 \times 1/3$$
$$V = 1{,}383.33 \text{ ft}^3$$

That answer might mess with your intuition for a second when you see that the number is smaller than before. Keep in mind that we are now talking about cubic feet instead of just square feet, so it isn't really a smaller amount of stuff any more than 1 gallon of water being less than 3 square inches of water. They aren't measuring the same thing. Square feet measures area in two dimensions, while cubic feet measures volume in three dimensions.

# Restaurant

Restaurant managers are responsible for the daily operations of their restaurants. Their oversight includes forecasting and ordering supplies and ingredients; employee recruitment; training; day-to-day management of kitchen, dining room, and support staff (sometimes with the aid of assistant managers); marketing; quality control; and customer relations. In most restaurant settings, managers will also be cross-trained in the various roles and will at times fill in as gaps between staffing and customer level warrant it. Additional responsibilities include ensuring that the facility and the staff meet various licensing requirements, addressing inspections and certifications, and serving as a **liaison** to community zoning and commerce agencies.

## Motion to table the issue

Brendan is thinking of buying some new tables for the bakery. He doesn't have lot of room, so he'd like to come up with the option that provides the most additional seating for a given amount of table area. The restaurant supply catalog has round, hexagonal, and rectangular tables. Even though they each say they can seat 6 people, he doesn't want his customers to feel crowded.

In order to see which one will actually be better for his customers, Brendan has to calculate the perimeter to help make his decision.

The perimeter of a circle is called the circumference ($C$), and the formula is:

$$C = 2\pi r$$

# TABLE OF TABLES

| Option | Round | Hexagonal | Rectangular |
|---|---|---|---|
| Seating | 6 people | 6 people | 6 people |
| Dimensions | 1.75-foot radius | 1.96 feet per side | 2.5 feet by 4 feet |
| Area | 10 ft² | 10 ft² | 10 ft² |

where:

$r$ = The radius of the circle.

We can find the perimeter of the first table:

$$C = 2\pi \times 1.75$$
$$C = 3.5\pi$$
$$C = 10.996 \text{ feet}$$

The perimeter of a hexagon ($P_h$) is:

$$P_h = 6a$$

where:

$a$ = The length of one side.

$$P_h = 6a$$
$$P_h = 6 \times 1.96$$
$$P_h = 11.76 \text{ feet}$$

Geometry in Others' Everyday Lives

And the perimeter of a rectangle ($P_r$) is:

$$P_r = 2l + 2w$$

where:

$l$ = Length of the rectangle.
$w$ = Width of the rectangle.

$$P_r = (2 \times 4) + (2 \times 2.5)$$
$$P_r = 8 + 5$$
$$P_r = 13 \text{ feet}$$

Therefore, the rectangular table offers the greatest perimeter, meaning the customers are least likely to feel crowded. Of course, Brendan would also have to consider the area of the tables, to make sure he could use his floor space efficiently.

## More parking

Let us imagine, as well, that our trusty, hard-working restaurant manager is thinking to have the parking lot outside repainted so as to improve traffic flow and, hopefully, to add some room for trees and landscaping along the central line of the lot. Here is a diagram of the current (gray) and the proposed (yellow lines) parking lot spaces.

The length and width of the current parking spaces are 9 feet and 18 feet. The new parking spaces will have the same dimensions, except the width of 9 feet is tilted in now to run perpendicular to the slanted parking lines. The 9-foot width is shown by the shortest leg of the red triangle in the diagram on page 111. This diagram shows using 30-60-90 triangles, common angles for parking lots. In addition to the spaces getting shorter (measured from left edge to right), the space *between* rows will shrink from 24 feet to 20 feet because people need less space to back out. How much space would be saved overall? Would that be enough to bring in a landscape architect to design a green space for the middle part of the lot?

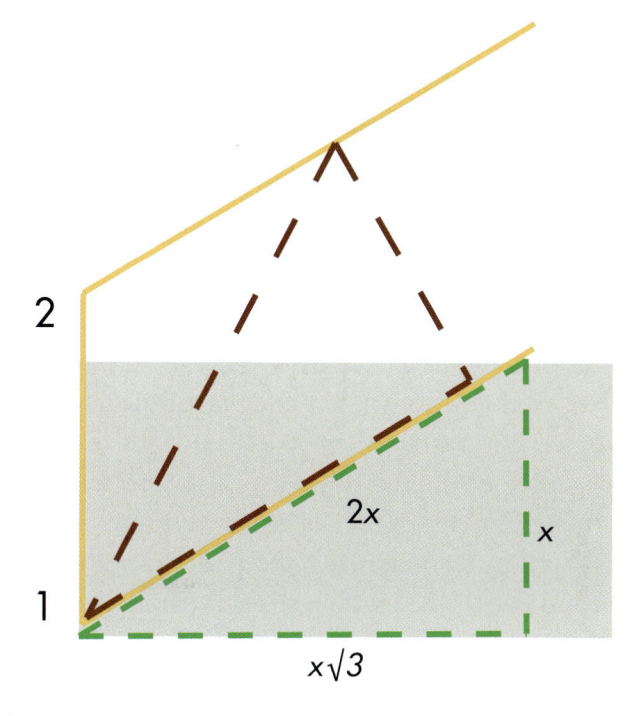

In order to find how much total width will be saved by resurfacing and repainting the lot, we need to find the length of the new parking spaces (the base of the green triangle). Fortunately we know the length of $x$, which is 9 feet. We also know the ratio of the sides, given that this is a 30-60-90 Pythagorean triangle. Now we just need to calculate by dropping 9 in for the $x$ on the bottom leg.

$$\text{length} = x\sqrt{3}$$
$$\text{length} = 9\sqrt{3}$$
$$= 9 \times 1.73205$$
$$= 15.58846 \text{ feet} \approx 15.6 \text{ feet}$$

Keep in mind we already know the length of the new parking lines (in yellow), which is the same as the hypotenuse of the green triangle:

$$2x = 2 \times 9 = 8 \text{ feet}$$

Are you with me? OK, good. There are two steps left to walk through. The first step is a nice trick you can run with 30-60-90 triangles. Notice that if we want a line perpendicular to the green hypotenuse, we just put another 30-60-90 right on top of it (in red). The short leg of the red triangle is also 9 feet. That will help guarantee that the width of each space is constant.

The last step in reconfiguring the parking spaces is to make sure the second line of the parking space is parallel to the first one. One cool way to do that is to use what we know from **axioms** of geometry. If the inside angle by the number 1 and the inside angle by the number 2 are supplementary (they add up to 180 degrees), then that shows us that the lines are parallel.

So, if there are currently four rows of parking, each 18 feet wide, and we have now cut each one down to 15.6 feet, that's a gain of 2.4 feet each.

$$2.4 \times 4 = 9.6 \text{ feet}$$

The reduced distance to safely back a car out saves 4 feet each. If the lot has 4 aisles between parking rows, that would give us:

$$(9.6 + 4) \times 4 = 25.6 \text{ feet}$$

That sounds like enough space to be able to green the place up, sure enough. Sweet.

# Physical Therapist

Physical therapists work with people to manage pain and recover strength and range of movement after illness or injury. Working closely with physicians, they are an important part of an overall health care team. They can also play a valuable role in patient education for general wellness and preventive health care.

The application of care involves a variety of hands-on therapy, like massage and assisted stretching, as well as the use of specialized medical and exercise equipment. They will provide careful measurement and documentation of **impairment**(s) and gradual improvement in the recovery of lost functions. Physical therapists also train the patient and/or family members for any specialized assistance or procedures needed at home to promote recovery.

## How do you measure up?

Much of the entire model of thinking about patient conditions revolves around lines, angles, and planes of movement. Geometry comes into play from the very beginning of care and treatment as the physical therapist uses a special series of tests and a device called a goniometer—think of a protractor for measuring angles but with two long arms sticking out from it.

The physical therapist will use visual estimation and actual measurement to document the injured body part in static positions through the possible planes along which it would normally be able to move or rotate. To get a sense of this, consider how the arm can bend or straighten at the elbow. This would constitute movement along one plane, with a normal healthy range of almost 180 degrees (fully extended to fully bent). Then consider that with your arm extended, you can move it from pointing straight out in front of you to pointing straight out to the side and continue on to point some amount toward the rear. This would constitute a second, horizontal, plane of movement. There is then the plane of movement that contains the arc from hand at your side up to

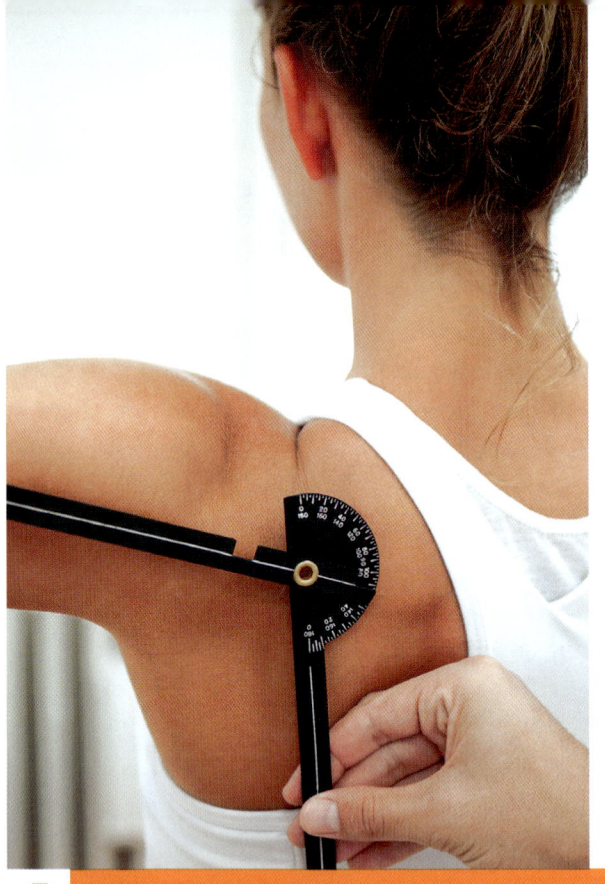

Physical therapists use a device to measure angles of flexibility and mobility in different parts of the body to help plan for improvements.

pointing straight out from the shoulder and then on up overhead—possibly on over beyond vertical. This is then compounded by the rotational capability of your arm while moving through any of these planes.

## Bad to the bone

Another interesting example comes up in the case of physical therapy adjustments after injuries or a surgery that affects the bones. The ability of a bone to stand up to tensile (bending) loads and torsion (twisting) is a factor of the thickness of the bone. Any injury or surgery which causes a decrease in the difference between the inner radius ($r_i$) and the outer radius ($r_o$) will result in a greater likelihood of fractures and breaks. That means exercise recommendations and weight amounts must be adjusted accordingly to prevent causing new injuries.

Gender can also affect the geometry involved. Female athletes, for instance, tend to have up to 3.5 times as many non-contact ACL (anterior cruciate ligament) injuries. This is one of the four major ligaments in the knee and its function is to help prevent side-to-side displacement of the knee. That is, it helps keep your shin from moving out of line with your thighbone when there is sideways pressure exerted.

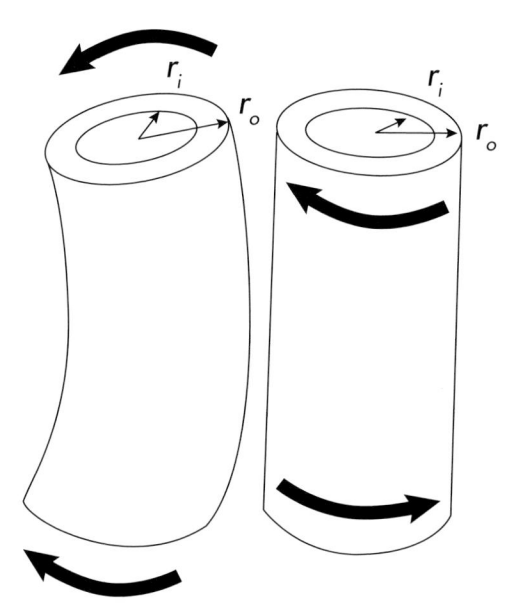

There are a number of factors that play into the higher injury rate, including the fact that females tend to have greater joint flexibility, which causes the knee to be less stable. It is also affected by the fact that because women have wider hips, the upper leg is coming into the knee joint at a greater angle, and that will change the stress load on the ligament. One of the ways to help in the conditioning process to prevent re-injury is to build strength in the hamstrings.

# CONCLUSION

There we have another epic adventure under our belts. We have taken a tour through the spaces large and small that make up our days, even if it was sometimes only the space in our heads as we tried to visualize the different jobs and aspects of our daily lives. We have gone from a ride at the amusement park to a bit of time working on the farm, and we have gone from a bit of sports to a nice sample of physical therapy to help us with rehab. (That one seems particularly fitting, don't you think?)

But of course, in the process, we have also done more than just wander through, I hope, staring on like passive spectators shuffling through some old museum. Rather, what I hope happened, and keeps happening, is for you to realize:

1. There are some puzzles out and about in the real world that math could help you solve in a shorter, lazier manner so you can get back to important things like keeping track of your favorite celebrity's gossip fight with your least favorite celebrity or trying to be the first person to prove you can gain IQ points by eating bacon.
2. Some math problems have a cool idea that they are built around that you might have fun thinking about even if you don't need to solve an actual problem.
3. There are few jobs that aren't able to benefit from a healthy interest and ability in

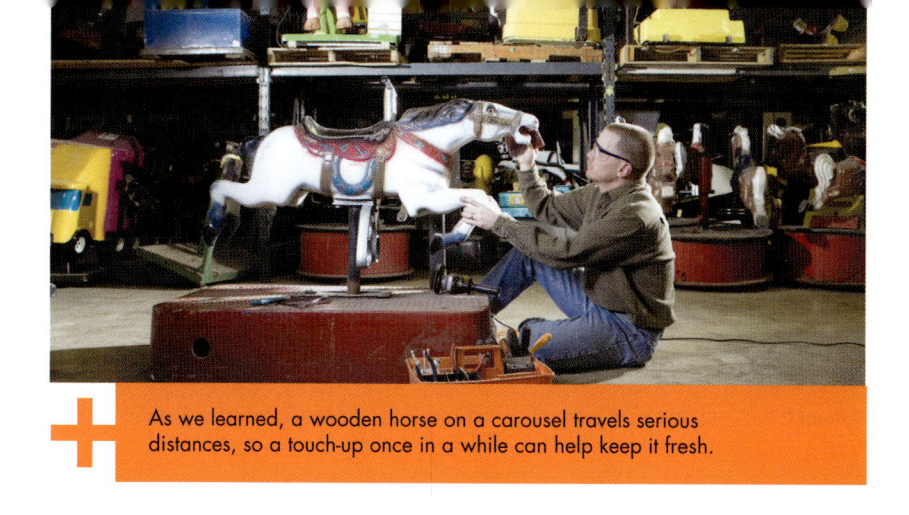

As we learned, a wooden horse on a carousel travels serious distances, so a touch-up once in a while can help keep it fresh.

different ways of looking at problems through the lens of mathematics—not just as a tool for putting numbers and measurements on everything, but also as a method of careful thinking.
4. Math is not just about making us better at work, as it might sometimes seem, but it is incredibly helpful at making us better at play!
5. Maybe math can be one of those things we do to play just for the sake of the coolness of seeing things in a new way.

Obviously some of those are meant a little more lightly than others. I hope you will come away with a couple of new or reconditioned tools in your mental toolbox to use on your quest. That's part of the beauty of including such a variety of activities and careers in the course of this book—not because we're making a guess about your interests or your likely career choice later on, but because there is such a high chance in your own life that you will come across challenging obstacles that are similar in key ways to some of these.

If that is not enough to help you see the value, then think of books like this and the geometry ideas in here as one of the things you pick up in a video game because they look like they might be useful later, even if you can't tell how yet. You never know when you might use them!

# Glossary

**acute** An angle with internal measurement less than 90 degrees.

**axiom** A statement or proposition considered as proven or established.

**caliph** The political and religious leader of a Muslim country.

**capacity** The maximum amount that something is able to hold.

**collaborative** Characterized by two or more people working together and sharing ideas.

**congruent** When two shapes have the same angle measurements and the same lengths of sides.

**deductive** Describing a line of reasoning or proof that proceeds from a general rule or axiom to a specific conclusion.

**elevation** How high something is.

**impairment** Being limited or reduced in some way compared to normal function.

**intrepid** Brave, fearless.

**isosceles** A triangle or trapezoid that has two sides of equal length.

**kendo** A sport based on a traditional Japanese style of sword fighting using two-handed bamboo swords.

**leverage** To gain advantage; to get extra work from something compared to what you seem to put in.

**liaison** A person who helps groups communicate and work more closely together.

**luminous** As if giving off light; bright.

**mandala** A symbolic geometric representation of the universe.

**medieval** Of or relating to the Middle Ages, which were roughly from the mid-fifth century to the mid-fifteenth century ce in the West.

**momentum** How much movement an object has; usually its mass times its speed.

**Moorish** Of or relating to the Moors, a Muslim culture from northwestern Africa.

**obtuse** An angle with internal measurement greater than 90 degrees.

**optimize** To see the maximum possible value for something.

**parallel** Lines or planes which are side by side but always remain an equal distance apart.

**perimeter** The distance around the outside edge of something.

**perpendicular** Oriented at 90 degrees from a given line or plane.

**plane** An imaginary flat surface defined by three points contained within it that are not all on the same line.

**point** A single location in space moving no distance in any direction.

**polyhedral** Characteristic of a geometric solid having multiple faces.

**postulate** A proposition or claim that is tentatively held as true in order to establish grounds for investigating some issue or concept.

**prism** A solid geometric figure with two congruent ends and rectangular sides.

**proportion** A relationship of equality between two ratios.

**Renaissance** A period of accelerated learning and development from roughly the fourteenth to the seventeenth centuries in the West.

**sprocket** The teeth sticking off the edge of a wheel in order to engage it with a chain drive; sometimes used to refer to the whole wheel itself.

**symmetry** Having congruent parts on opposite sides of some axis.

**Tibet** A small country in the Himalayas to the northwest of China; it has been under Chinese control since China invaded Tibet in 1950.

**trapezoid** A quadrilateral with one, and only one, pair of parallel sides.

**trigonometry** A branch of math that focuses on properties and relations among sides and angles of triangles. It also works with cyclical functions.

**variables** Parts of a formula or equation that can stand in for different possible values.

**vertex** The point of an angle or polygon where two or more lines come together.

**Zen** A Japanese form of religion that focuses on meditation and compassionate action.

# Further Reading

## Books

Gowers, Timothy, June Barrow-Green, and Imre Leader, eds. *The Princeton Companion to Mathematics*. Princeton, NJ: Princeton University Press, 2003.

Huettenmueller, Rhonda. *Precalculus Demystified: Hard Stuff Made Easy*. New York: McGraw-Hill Education, 2012.

Jackson, Tom, ed. *Mathematics: An Illustrated History of Numbers*. New York: Shelter Harbor Press, 2012.

Pickover, Clifford A. *The Math Book: From Pythagoras to the 57th Dimension, 250 Milestones in the History of Mathematics*. New York: Sterling, 2009.

Simmons, George F. *Precalculus Mathematics in a Nutshell: Geometry, Algebra, Trigonometry*. Eugene, OR: Wipf and Stock, 2003.

## Websites

**Agnes Scott College**
www.agnesscott.edu/lriddle/women/chronol.htm

This site has a great collection of biographies of important women mathematicians down through history.

**Cool Math**
www.coolmath.com/precalculus-review-calculus-intro

This is a good website for a wide range of topics. Of particular note is the colorful layout for a good review of precalculus material.

### Drexel University Math Forum
mathforum.org/dr.math

This site offers a variety of basic explanations of concepts and examples. One of its strengths is that the topics are broken out by subjects and by level in school (elementary, middle, high school, college and beyond). The site does a helpful job of working through example problems and also allows you to search by topic.

### Get the Math
www.thirteen.org/get-the-math

This site has some good videos geared at the middle-school level that give interaction with real-life examples from fashion, music, and more.

### Khan Academy
www.khanacademy.org

This is an excellent online learning website, and the learn-as-you-go format allows you to click on only the short instructional videos you need as you work your way through the course. There are courses at a wide range of skill levels.

### NYU Department of Mathematics—Courant Institute
cims.nyu.edu/~kiryl/precalculus.html

This is a nice, clean, visual layout of a thorough variety of mathematical topics, as well as review of key ideas from algebra.

***Plus* Magazine**
plus.maths.org/content/

This online magazine is run under the Millennium Mathematics Project at Cambridge University. It offers interesting short articles on interesting problems in math (and science). The problems are complex, but they are broken down in a way that is easy to follow.

# Bibliography

Bellos, Alex. *The Grapes of Math*. New York: Simon and Schuster, 2014.

Berlinghoff, William P., Kerry E. Grant, and Dale Skrien. *A Mathematical Sampler: Topics for Liberal Arts*. Lanham, MD: Ardsley House Publishers, 2001.

Byers, William. *How Mathematicians Think: Using Ambiguity, Contradiction, and Paradox to Create Mathematics*. Princeton, NJ: Princeton University Press, 2007.

Clawson, Calvin C. *Mathematical Mysteries: The Beauty and Magic of Numbers*. New York: Plenum Press, 1996.

Devlin, Keith. *Mathematics: The Science of Patterns: The Search for Order in Life, Mind and the Universe*. New York: Henry Holt, 2003.

Ellenberg, Jordan. *How Not to Be Wrong: The Power of Mathematical Thinking*. New York: Penguin, 2014.

Freeman, W. H., Brendan Cady, ed. *For All Practical Purposes: Mathematical Literacy in Today's World*. New York: W. H. Freeman and Company, 2006.

Frenkel, Edward. *Love & Math: The Heart of Hidden Reality*. New York: Basic Books, 2013.

Gordon, John N., Ralph V. McGrew, and Raymond A. Serway. *Physics for Scientists and Engineers*. Belmont, CA: Thomson Brooks/Cole, 2005.

Jackson, Tom, ed. *Mathematics: An Illustrated History of Numbers*. New York: Shelter Harbor Press, 2012.

Kanold, Timothy. *Geometry: Integration, Applications, Connections*. Columbus, OH: McGraw-Hill, 2001

Lehrman, Robert L. *Physics the Easy Way*. Hauppage, NY: Barron's Educational Series, Inc., 1990.

Mankiewicz, Richard. *The Story of Mathematics*. Princeton, NJ: Princeton University Press, 2000.

McLeish, John. *Numbers: The History of Numbers and How They Shape Our Lives*. New York: Fawcett Columbine, 1991.

Oakley, Barber. *A Mind for Numbers: How to Excel at Math and Science (Even if You Flunked Algebra)*. New York: Tarcher, 2014

Paulos, John Allen. *Beyond Numeracy: Ruminations of a Numbers Man*. New York: Alfred A. Knopf, 1991.

Peterson, Ivars. *Islands of Truth: A Mathematical Mystery Cruise*. New York: W. H. Freeman and Company, 1990.

———. *The Mathematical Tourist: Snapshots of Modern Mathematics*. New York: W. H. Freeman and Company, 1988.

Pickover, Clifford A. *The Math Book: From Pythagoras to the 57th Dimension, 250 Milestones in the History of Mathematics*. New York: Sterling, 2009.

Rooney, Anne. *The Story of Mathematics: From Creating the Pyramids to Exploring Infinity*. London: Arcturus, 2015.

Sardar, Ziauddin, Jerry Ravetz, and Borin Van Loon. *Introducing Mathematics*. Cambridge, UK: Icon Books, 1999.

Serra, Michael. *Discovering Geometry: an Inductive Approach*. Oakland, CA: Key Curriculum Press, 1997.

Stewart, Ian. *Nature's Numbers: The Unreal Reality of Mathematics*. New York: Basic Books, 1995.

Stewart, James, Lothar Redlin, and Saleem Watson. *Elementary Functions*. Boston: Cengage Learning, 2011.

Strogatz, Steven. *The Joy of X: A Guided Tour of Math, from One to Infinity*. Boston: Houghton-Mifflin, 2012.

# Index

Page numbers in **boldface** are illustrations. Entries in **boldface** are glossary terms.

**acute**, 56
amusement parks, 19–24
archaeology, 72–82
**axiom**, 112

backpacking, 47–50
basketball, 46–47
bicycles, 38–41, **38**, **41**, 55–56, **55**
boats, 50–52

**caliph**, 13
**capacity**, 50
carousels, **18**, 21–24, **117**
**collaborative**, 83
**congruent**, 29, 52
cubism, 24–25, **25**

**deductive**, 9–10

**elevation**, 79
Escher, M. C., 28, 33
Euclid, 10–11, **10**, 14, 17

farming, 87–91, **88**, 94–101

goniometer, 113, **114**

hang gliding, 56–58
heat sinks, 41–46, **43**

**impairment**, 113
interior design, 83–86, **85**, 92–93
**intrepid**, 76
**isosceles**, 56

keel, 51–52, **51**
**kendo**, 66, **67**

landscape architecture, 101–108, 111
**leverage**, 79
**liaison**, 108
**luminous**, 17

**mandala**, 72, 74–78, **74**, **76**, 80
**medieval**, 23
Millay, Edna St. Vincent, **16**, 17
**momentum**, 56
**Moorish**, 33
mound builders, 80–82, **81**

126  Applying Geometry to Everyday Life

nautical miles, 63
non-Euclidean geometry, 11, 14–15

**obtuse**, 56
**optimize**, 83

**parallel**, 14–15, 52, 83–84, 91–92, 94, 112
parking lots, 110–112, **111**
**perimeter**, 64, 90, 98, 108–110
perpendicular, 83–84, 94, 110, 112
perspective, 11, **11**, 14, 24–25, 33
physical therapy, 71–72, 113–115, **114**
Picasso, Pablo, 24–25
**plane**, 9, 14, 19, 25, 29, 32–33, 35, 66, 69, 72, 113–114
Plato, 9–10, **9**
**point**, 24, 49, 51–52, 58
pointillism, 24, **25**
**polyhedral**, 35
**postulate**, 14, 52
**prism**, 65–65, **88**, 97, 104
**proportion**, 11, 22, 24, 39, 61, 82, 87
Pythagorean theorem, 8–10, 44, 48, 59, 61, 78, 91, 111

quilts, 26–28, **26, 28**, 30–31

racetracks, 52–54, **53**
**Renaissance**, 7, 11, 14
restaurant management, 108–110

sculpture, 34–37, **35**
Seurat, Georges, 24, **25**
skateboarding, 67, **68**, 69
**sprocket**, 38–40, **38**
swimming pools, 64–66, **64**
**symmetry**, 24, 30, 56, 59, 66, 72–73, **73**

tennis, 58–61, **58, 60**
tessellation, 28–29, **28, 29**, 32–34
**Tibet**, 72, 74
**trapezoid**, 29, 51, 82, 97, 100
**trigonometry**, 11, 37

**variables**, 48, 52, 55
**vertex**, 29, 32, 57, 90

Zen rock gardens, **70**, 71–72, 104–108

# About the Author

**Erik Richardson** is an award-winning teacher from Milwaukee, where he has taught and tutored math up to the college level over the last ten years. He has done graduate work in math, economics, and the philosophy of math, and he uses all three in his work as a business consultant with corporations and small businesses. He is a member of the Kappa Mu Epsilon math honor society, and some of his work applying math to different kinds of problems has shown up at conferences, in magazines, and even in a few pieces of published poetry. As the director of Every Einstein (everyeinstein.org), he works actively to get math and science resources into the hands of teachers and students all over the country.